Yair Linn

Synchronization in Coherent M-PSK Receivers

Yair Linn

Synchronization in Coherent M-PSK Receivers

Carrier Synchronization, Phase Detection, Lock Detection, and SNR Estimation

VDM Verlag Dr. Müller

Impressum/Imprint (nur für Deutschland/ only for Germany)
Bibliografische Information der Deutschen Nationalbibliothek: Die Deutsche Nationalbibliothek
verzeichnet diese Publikation in der Deutschen Nationalbibliografie; detaillierte bibliografische
Daten sind im Internet über http://dnb.d-nb.de abrufbar.
Alle in diesem Buch genannten Marken und Produktnamen unterliegen warenzeichen-, marken-
oder patentrechtlichem Schutz bzw. sind Warenzeichen oder eingetragene Warenzeichen der
jeweiligen Inhaber. Die Wiedergabe von Marken, Produktnamen, Gebrauchsnamen,
Handelsnamen, Warenbezeichnungen u.s.w. in diesem Werk berechtigt auch ohne besondere
Kennzeichnung nicht zu der Annahme, dass solche Namen im Sinne der Warenzeichen- und
Markenschutzgesetzgebung als frei zu betrachten wären und daher von jedermann benutzt
werden dürften.

Coverbild: www.purestockx.com

Verlag: VDM Verlag Dr. Müller Aktiengesellschaft & Co. KG
Dudweiler Landstr. 125 a, 66123 Saarbrücken, Deutschland
Telefon +49 681 9100-698, Telefax +49 681 9100-988, Email: info@vdm-verlag.de
Zugl.: Vancouver, University of British Columbia, Diss., 2007

Herstellung in Deutschland:
Schaltungsdienst Lange o.H.G., Zehrensdorfer Str. 11, D-12277 Berlin
Books on Demand GmbH, Gutenbergring 53, D-22848 Norderstedt
Reha GmbH, Dudweiler Landstr. 99, D- 66123 Saarbrücken
ISBN: 978-3-639-08144-2

Imprint (only for USA, GB)
Bibliographic information published by the Deutsche Nationalbibliothek: The Deutsche
Nationalbibliothek lists this publication in the Deutsche Nationalbibliografie; detailed
bibliographic data are available in the Internet at http://dnb.d-nb.de.
Any brand names and product names mentioned in this book are subject to trademark, brand or
patent protection and are trademarks or registered trademarks of their respective holders. The use
of brand names, product names, common names, trade names, product descriptions etc. even
without
a particular marking in this works is in no way to be construed to mean that such names may be
regarded as unrestricted in respect of trademark and brand protection legislation and could thus
be used by anyone.

Cover image: www.purestockx.com

Publisher:
VDM Verlag Dr. Müller Aktiengesellschaft & Co. KG
Dudweiler Landstr. 125 a, 66123 Saarbrücken, Germany
Phone +49 681 9100-698, Fax +49 681 9100-988, Email: info@vdm-verlag.de

Produced in USA and UK by:
Lightning Source Inc., 1246 Heil Quaker Blvd., La Vergne, TN 37086, USA
Lightning Source UK Ltd., Chapter House, Pitfield, Kiln Farm, Milton Keynes, MK11 3LW, GB
BookSurge, 7290 B. Investment Drive, North Charleston, SC 29418, USA
ISBN: 978-3-639-08144-2

TABLE OF CONTENTS

LIST OF TABLES

LIST OF FIGURES

LIST OF ABBREVIATIONS

AGC	Automatic Gain Control
ASIC	Application Specific Integrated Circuit
BPSK	Binary Phase Shift Keying
CRB	Cramér-Rao Bound
DA	Data Aided
DD	Decision Directed
DDS	Direct Digital Synthesizer
D-MPSK	Differential M-ary Phase Shift Keying
DVB	Digital Video Broadcasting
ECD	Error Correction Decoder
FFT	Fast Fourier Transform
FPGA	Field Programmable Gate Array
IAD	Integrate and Dump
IF	Intermediate Frequency
LSB	Least Significant Bit
LUT	Lookup Table
M-PSK	M-ary Phase Shift Keying
MSB	Most Significant Bit
MSE	Mean Squared Error
NCO	Numerically Controlled Oscillator
NDA	Non-Data Aided
NMSE	Normalized Mean Squared Error
PD	Phase Detector
PLL	Phase Locked Loop
PSK	Phase Shift Keying
QPSK	Quaternary Phase Shift Keying
RF	Radio Frequency

RMS	Root Mean Square
SER	Symbol Error Rate
SNR	Signal-to-Noise Ratio
SRRC	Square Root Raised Cosine
SVR	Signal-to-Variation Ratio
TED	Timing Error Detector
VCO	Voltage Controlled Oscillator
WPAN	Wireless Personal Area Network

ACKNOWLEDGEMENTS

Where do I begin? First and foremost I would like to thank my former supervisor at UBC, Prof. Matthew J. Yedlin, for his kindness, gentle guiding, wisdom and support throughout my thesis writing process and indeed throughout my studies at UBC. No words would be adequate to express my gratitude, and so I might as well stop here.

A mentor of mine since my early days as an undergraduate is Prof. Shmuel Zaks, of the Technion Israel Institute of Technology. I have known him for more than 15 years, and throughout all this time he has provided me with guidance, wisdom, and assistance that I found invaluable and indispensable.

Prof. Robert Schober has also been a friend as well as a purveyor of professional advice throughout my studies, for which I thank him profusely. I would also like to thank him for serving as my supervisor at UBC as Prof. Yedlin's proxy while the latter was on sabbatical, and for handling the bureaucratic tasks associated with both my departmental and final defense. His selfless contribution my thesis writing process is much appreciated. I would also like to thank my third supervisory committee member, Prof. Steve Wilton, for his guidance and for his excellent course on FPGAs.

Prof. Sayra M. Cristancho and Prof. Alex A. Monclou from the Universidad Pontificia Bolivariana in Bucaramanga, Colombia, have been great friends throughout these past few years and have provided me with personal and professional assistance for which I am particularly grateful and for which I will be eternally in their debt.

Last but by no means least, I would like thank my family for their help. My parents, Shai and Ruth, for their financial support and encouragement. My aunt Orna for various matters for which her help was indispensable. My sister, Gilat, and her husband Peter, for their support in the most crucial moments, and my brother Erez for various errands he performed on my behalf. Lastly, I would like to thank my grandparents Ruth and Amnon, who have always served as an inspiration to me throughout my life and my studies.

To my sister, Gilat,

for being there when I needed her most

Chapter 1   Introduction and System
Model Definitions

Digital wireless communications systems involve three general elements: the transmitter, the channel, and the receiver. The transmitter's purpose is to modulate the data stream and send the RF (Radio Frequency) signal over the channel, where that signal is corrupted by noise and possibly also interference and distortion. The receiver's purpose is to recover the original data stream while overcoming to the best of its ability the malicious effects of noise, interference, and distortion. This process is known as demodulation.

In this book we shall investigate structures for coherent demodulation of single-carrier M-PSK signals. M-PSK modulation is used in a wide variety of contemporary broadcasting and networking standards. For example, M-PSK is used in various configurations of the DVB-S [2], DVB-S2 [3], DVB-T [4], DVB-H [5], WiFi (802.11) [6], WPAN (802.15) [7], and WiMAX (802.16) [8] standards, to name a few. M-PSK also has many applications in military communications, especially using satellites [9], [10], [11].

Design of coherent receivers in digital communications involves generating a local carrier that is in phase with the received carrier, and then using this local carrier in order to demodulate the received signal. Generation of the local carrier can be done in one of the following manners: (a) by using pilot symbols or pilot signals, or (b) by extracting of carrier phase information from the received signal itself. The use of pilot signals or pilots symbols has the distinct and inevitable disadvantage of necessitating the expenditure of transmitter power that could otherwise have been used to increase the transmission power of the information-bearing signal (or, alternatively, to increase the the data rate). Hence, if possible, one would like to avoid the use of pilot signals or symbols and endeavour to regenerate a local carrier using only information obtainable from the information-bearing received signal. This indeed shall be the focus of this book.

In coherent suppressed-carrier receivers, regeneration of the local carrier is generally done via a Phase Locked Loop (PLL) that operates on the output of a Phase Detector

(PD). The phase detector provides an indication of the residual phase error between the local and received carriers, and the PLL acts in feedback in order to cancel that phase error.

Recovery of the symbol clock is also done using a PLL, which in this instance is called the symbol timing synchronization PLL (or symbol PLL for short). The purpose of the symbol PLL is to generate a local symbol clock that is time-coherent with the received signal's symbol clock. This allows the receiver to sample the received symbol waveforms at the optimal times. The symbol PLL operates upon an error signal that is supplied by a Timing Error Detector (TED).

Another essential element in any PLL circuit is a lock detector. In a carrier PLL the purpose of the lock detector is to indicate when the local carrier is phase-coherent with the received carrier, in which case the carrier PLL is deemed "locked". Timely lock detection is necessary for many receiver operations. For example, when the carrier PLL becomes locked the receiver needs to stop scanning the input frequency uncertainty region in order to avoid driving the carrier PLL out of lock.

A crucial element in any modern communications receiver is an SNR (Signal-to-Noise Ratio) estimation module. In many modern communications schemes an accurate E_s/N_0 estimate is needed not only as a monitoring aid, but rather it plays an important role in the receiver's operation. For example, some error correction decoders can make use of an E_s/N_0 estimate to increase their coding gain (e.g. turbo codes [12]). Another example are systems that employ diversity reception, for which SNR estimates are used to assign relative weights to the data obtained from the various receivers [13 Sec. 14.4]. As another example we mention adaptive schemes (e.g. [14], [15]) where the data and/or coding rates are altered according to the E_s/N_0. See also [16 Sec. 1.2] for a more extensive overview of the uses of SNR estimators in contemporary communications systems, along with many useful references. In this book, we shall present new SNR estimators for M-PSK and D-MPSK receivers.

1.1   Overview of This Book

In the remainder of this introductory chapter (Chapter 1) we shall provide an overview of the signal and receiver models pertaining to this book. The object of this book is to present new structures for M-PSK lock detection (Chapter 2), phase detectors for coherent M-PSK carrier synchronization (Chapter 3), and SNR estimation in M-PSK and D-MPSK receivers (Chapter 4). The final chapter (Chapter 5) is devoted to conclusions, and is followed by several appendices which contain important mathematical derivations which were relegated there in order to maintain the book's flow.

As we shall see, the structures presented in this book are interrelated, and often one can obtain added benefit by exploiting these interrelationships. Although the digital portion of the receiver could be implemented either in hardware or in software, this book focuses on fixed-point hardware implementations. The reason is that, first, fixed-point hardware implementations will always have distinct performance advantages due simply to the fact that fixed-point hardware implementations can always be made to operate faster than any software and/or floating point implementation ([17 Chap. 9], [18], [1]). Secondly, more intriguing challenges are present when trying to design a hardware system: while a software system could be implemented in a high-level language, in contrast when implementing a fixed-point hardware system the designer must explicitly address such issues as scaling and dynamic range, logic resource usage, implementation of mathematical operations, etcetera. Finally, given the preponderance of FPGA-based and ASIC-based receivers, such a focus on fixed-point hardware implementations is appropriate in the context of contemporary trends in receiver design.

1.2 Intuitive Understanding of M-PSK Synchronization

Before delving into mathematics, it may be worthwhile to attempt to attain an intuitive understanding of the meaning of carrier and symbol synchronization in an M-PSK receiver. To this end, let us assume a BPSK communications system where the baseband data pulse is rectangular, and let us look at the waveforms at the outputs of the I and Q matched filters (note that the post-matched-filter pulse shape is *triangular* [13 Sec. 5.1]). In the following figures, the system's symbol rate is denoted as $1/T$.

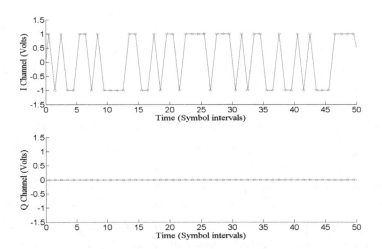

Fig. 1. Post-matched-filter signals. The received signal is noiseless BPSK. No carrier error, no timing error.

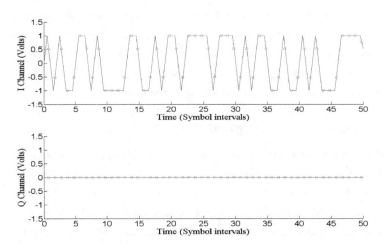

Fig. 2. Post-matched-filter signals. The received signal is noiseless BPSK. No Carrier error. Timing error of $T/4$.

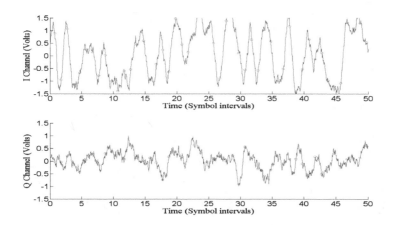

Fig. 3. Post-matched-filter signals. The received signal is BPSK with $E_S / N_0 = 7\,\mathrm{dB}$. No Carrier error. Timing Error of $T/4$.

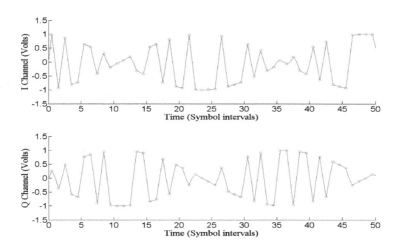

Fig. 4. Post-matched-filter signals. The received signal is noiseless BPSK. Carrier frequency error of $1/(50 \cdot T)$. No timing error.

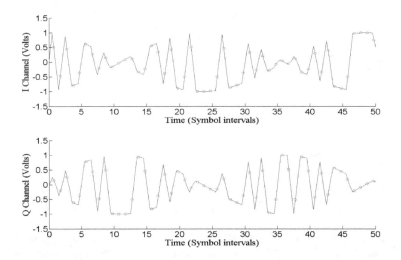

Fig. 5. Post-matched-filter signals. The received signal is noiseless BPSK. Carrier frequency error of $1/(50 \cdot T)$ **. Timing error of** $T/4$ **.**

In Fig. 1 we see the situation that occurs when there is perfect carrier and symbol synchronization (a noiseless BPSK signal is assumed). The small circles denote the samples of the I and Q channels (at rate $1/T$, i.e. 1 sample/symbol, which is the rate needed by the carrier PLL). As we can see, the receiver's carrier is in complete synchronization with the input carrier (as seen by the fact that the Q channel is always 0) and the symbol timing synchronization loop is also working perfectly (as seen by the fact that the samples of the I channel are always taken at the symbol's peak).

In Fig. 2 we can observe the effects of a symbol timing synchronization mismatch between the local and received symbol clocks. The lack of symbol timing synchronization is evident in that the samples (the small circles) are now not taken at the peak of the symbols but rather at an offset of $T/4$ seconds. Though these sampling instances are not ideal, in the case of a noiseless BPSK signal they will have no effect upon the error rate since the data decision algorithm, which in this case is simply *sign(I)*, is unaffected. However, there will be a very appreciable impact upon the error rate once noise effects are taken into account. To see this, observe Fig. 3, which shows the effects

-6-

of a timing error of $T/4$ upon the reception of a BPSK signal with $E_s / N_0 = 7$ dB. In this case the error rate of the decision algorithm *sign(I)* is significantly degraded due to the timing error.

Now let us take a look at the effects of carrier frequency errors. In Fig. 4 we see the case where there is a carrier frequency error (but no timing error). The carrier frequency error causes the signal to meander between the I and Q channels. In Fig. 5 we see what happens when the carrier loop is unlocked and there is also a timing mismatch between the receiver and transmitter symbol clocks. The carrier frequency error manifests itself in the signal meandering between the I and Q channels, while the timing error manifests itself in that the samples (the small circles) are offset from the peaks of the symbols. Clearly, even for the noiseless BPSK signal shown in Fig. 4 or Fig. 5, there is an extraordinary degradation in the error rate as a result of the carrier frequency error, and when noise effects are taken into account there will be an even more pronounced degradation that will also be adversely affected by the lack of symbol synchronization.

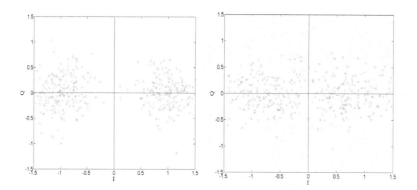

Fig. 6. I-Q Graphs for $E_s / N_0 = 7$ dB**. Left: no timing or carrier error. Right: no carrier error, timing error of** $T/4$**.**

It is also instructive to look at various scenarios using I-Q graphs. In Fig. 6 we see samples of a BPSK signal where the carrier PLL is locked. On the left-hand side of Fig. 6 we see a signal with $E_s / N_0 = 7$ dB where the symbol timing recovery is perfect. On the right, we see the effects of a timing error of $T/4$. Clearly, the timing error causes some data points to cross the decision region boundary (which is the vertical line $I=0$), hence worsening the error rate as compared to the perfect-timing recovery case.

In Fig. 7's left we see the effect of carrier frequency error on the I-Q graph; obviously, in this case no data recovery is possible. The case of a constant carrier phase error is shown on the right-hand side. Though data recovery is possible, the error rate will suffer tremendously as a result of the fact that many more of recovered symbols have crossed the decision region boundary $I=0$ (as compared to Fig. 6 left).

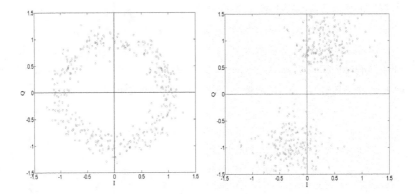

Fig. 7. I-Q Graphs. Left: $E_s/N_0 = 15$ dB with carrier frequency error, no timing error. Right: $E_s/N_0 = 7$ dB with constant carrier phase error of 0.4π, no timing error.

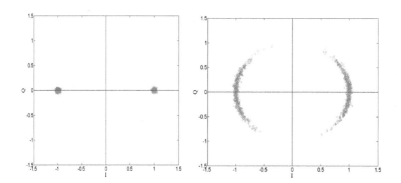

Fig. 8. *I-Q* graphs for BPSK with $E_S/N_0 = 30\,\mathrm{dB}$. Left: no carrier or timing error. Right: Gaussian carrier phase jitter with $\sqrt{\mathrm{var}(\theta_e)} = 20^\circ$, no timing error.

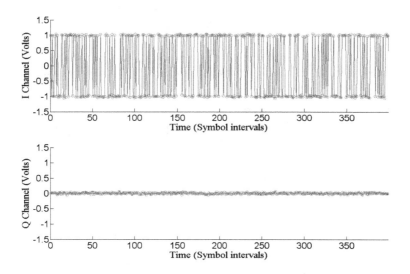

Fig. 9. Post-matched-filter signals. The received signal is BPSK with $E_S/N_0 = 30\,\mathrm{dB}$. No Carrier error, no timing error.

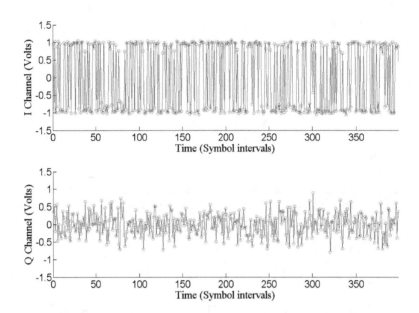

Fig. 10. Post-matched-filter signals. The received signal is BPSK with $E_S/N_0 = 30\,\text{dB}$. Gaussian carrier phase jitter with $\sqrt{\text{var}(\theta_e)} = 20^\circ$, no timing error.

In Fig. 8 we see the effects of carrier phase jitter. On the left, we see how an *I-Q* graph would look for $E_S/N_0 = 30$ dB for an ideally synchronized receiver. On the right, we see the effects of carrier phase jitter, that is a phase error θ_e which has a zero-mean and Gaussian distribution with $\sqrt{\text{var}(\theta_e)} = 20^\circ$. The corresponding *I* and *Q* channel graphs are shown in Fig. 9 and Fig. 10.

Although for BPSK and $E_S/N_0 = 30$ dB the effects of the carrier phase jitter shown in Fig. 8 (right) and Fig. 10 upon the error rate will be mild, the effect upon the error rate would be much more grave for a lower E_S/N_0 and/or a higher modulation index (e.g. QPSK, 8-PSK, etc.). We can see this in examining the *I-Q* graph of such a situation as shown in Fig. 11. As seen there, the phase jitter in this case compromises the separation between the received constellation signal points to such an extent that it causes the error rate to be substantially increased.

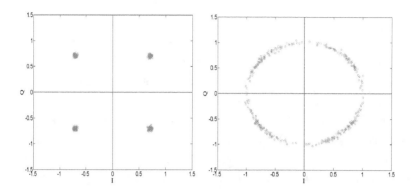

Fig. 11. I-Q graphs for QPSK with $E_S / N_0 = 30$ dB. Left: no carrier or timing error. Right: Gaussian carrier phase jitter with $\sqrt{\operatorname{var}(\theta_e)} = 20^o$, no timing error.

Clearly, as we've seen in this section, achieving carrier and symbol timing synchronization is crucial for proper data recovery. In this section we proceeded with the aim of attaining an intuitive (and, hence, qualitative) understanding of the meaning of synchronization. Quantitative effects of carrier phase jitter and symbol timing error upon the error rate of M-PSK can be found, for example, in [19 Sec. 4.3], [20], [21 Sec. 2.2.5], [22 Chap. 7], [23], and [24].

In the receiver, the synchronized carrier and symbol clock are generated by using PLLs which may have various topologies, an issue that is discussed in the next section.

1.3 Coherent M-PSK Receiver Topologies

There are three types of receiver topologies: (a) analog; (b) digital; and (c) hybrid. In the following three subsections, these architectures are discussed.

1.3.1   Analog architectures

In this case the PLL is implemented using analog components. Examples can be found in many texts, e.g. [25 Chap. 11], [13 Chap. 6], [26], [27 Chap. 10], [28 Chap. 11], [29 Chap. 9], [11]. Analog implementations suffer from some inherent problems, including: (a) variation of system parameters due to component value fluctuations; (b) parasitic and secondary effects which cause degradations; and (c) crosstalk-induced performance degradation. These problems are eliminated or at least significantly mitigated when using a digital implementation or a hybrid implementation.

Until the mid-to-late 1980's to the beginning of the 1990's, analog implementations were the workhorses of demodulators for high-speed communications. With the advent of powerful and cheap microelectronic circuits over the last 2 decades (including, notably, high-speed samplers and fast and dense FPGAs), purely analog implementations are quickly being abandoned in favour of digital and hybrid systems.

1.3.2   Digital architectures

Digital receiver architectures have been the subject of much investigation over the past 20 years. Two general subclasses of this architecture are possible.

a)   IF sampling

In the IF-sampling topology, the IF (Intermediate Frequency) signal is sampled, and the downconversion and demodulation is performed digitally on the sampled IF signal. Since the IF frequency is usually much higher than the symbol rate, this means that the sampling of the IF signal must be done at a rate much higher than the theoretical minimum of 2 samples/symbol, which renders this approach impractical for demodulating high datarate signals.

Sometimes the IF signal can be sampled using bandpass sampling techniques (see for example [30 Sec. 2.3.2], [31 Sec. 10.4.3], [17 Chaps. 7, 10]). However, even with this approach the sampling rate is usually much higher than 2 samples/symbol, particularly when one takes into account the necessity to increase the sampling rate in order to be able to handle carrier frequency uncertainties.

Thus, due to the high sampling rate required as compared to the symbol rate, the IF-sampling topology is usually adopted only for demodulation of low datarate communications.

b)   Near-baseband sampling

In the near-baseband topology, coarse frequency downconversion is done in the analog domain and the resulting near-baseband signal is sampled [22 Chap. 4]. Fine downconversion and demodulation is then performed entirely in the digital domain upon these near-baseband samples. Perhaps the most comprehensive treatment of this subject can be found in [22] (see an overview of this architecture in [22 Chap. 4]). An excellent tutorial and many important results can be found in [32] and [33].

The near-baseband digital architecture offers the important advantage of considerably simplifying the analog section of the receiver (essentially reducing it to an AGC (Automatic Gain Control) circuit followed by a coarse downconversion circuit). However, the digital implementation has two big drawbacks. First, the need to perform downconversion and interpolation ([32], [33], [22]) in the digital domain implies a rather complicated digital section. Secondly, the sampling rate necessary for good performance is at least 2 samples/symbol, and more likely at least 2.5 to 3 samples/symbol (see [33 Tables IV, V]). This, as compared to a minimum of 1 sample/symbol that is necessary for a hybrid implementation (see Sec. 1.3.3).

As sampler and FPGA/ASIC speeds become higher and digital logic densities increase, there is no doubt that digital implementations will gradually replace hybrid and analog implementations for many communications systems. However, due to the high sampling rate requirement and the high digital logic complexity, there shall always be a sizeable portion of receivers which are implemented using hybrid architectures, especially for high datarate communications.

1.3.3   Hybrid architectures

In a hybrid architecture, some of the PLL components are implemented in the analog domain while others are implemented using digital logic. Since many components can be implemented either analogically or digitally, this gives rise to many architectural possibilities. For example, in one hybrid implementation of a carrier PLL only the phase detector is implemented digitally, while in another implementation the digital portion would also include the loop filter and the matched filters. In general, the choice as to whether to implement a certain component in the analog or digital domains will come down to economics: i.e. what is the cheapest and/or easiest and/or most performance-effective way to implement that particular component, or what is the best implementational tradeoff given these constraints. An example hybrid implementation of an M-PSK receiver is shown in Fig. 12, although the reader is advised to remember that other variations on this architecture are possible.

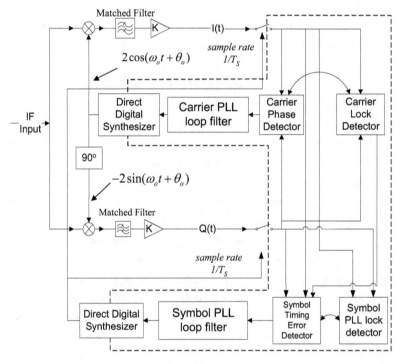

Fig. 12. General structure of a hybrid receiver for digital communications. The parts within the dashed line are implemented digitally, while the rest are analog components (the samplers and DDS (Direct Digital Synthesizer) are mixed-signal components).

Hybrid implementations offer the important advantages of being able to operate with excellent performance at sample rates as low as 1 sample/symbol (if a 1 sample/symbol TED is used, such as in various detectors discussed in [34], [35], [36], [21 Chaps. 7, 8] and [37], [38], [39], [40]). This is at least 2 to 3 times lower than the necessary sampling rate for a digital topology [33]. Moreover, the digital logic can be made to be extremely simple (see [41], [18], [1]). At a rate of 2 samples/symbol, virtually ideal performance can be achieved and carrier-independent timing error detectors that require 2 samples/symbol can be used ([42], [43], [37], [44]). The rate of 2 samples/symbol is still 25%-50% lower than the 2.5 to 3 samples/symbol usually needed for using a simple linear interpolator within the digital topology [33 Table V]. The hybrid topology is thus particularly suited for high-datarate communications (for example (in current technology) for symbol rates which are above 50 MHz) where sampling and real-time processing of the IF or near-baseband signal is often either impossible or uneconomical.

Using the hybrid architecture allows the designer to enjoy architectural benefits which are unavailable when using a completely analog or completely digital receiver. As alluded to earlier, sampling the baseband signal rather than the IF or near-baseband signal generally allows more inexpensive samplers which operate at a lower clock rate to be used. Furthermore, the fact that downconversion of the incoming signal is done in the analog domain considerably simplifies the digital logic required, since the latter is relieved from the need to perform downconversion of the complex (i.e. I-Q) sampled signal.

On the other hand, having the I-Q demodulator in the analog domain driven by an analog local carrier allows an arbitrarily high IF frequency f_{IF} to be used with the aid of an external mixer and a fixed oscillator. Thus, with a relatively inexpensive low frequency DDS (Direct Digital Synthesizer) and an additional (relatively inexpensive) fixed RF oscillator, the use of an expensive and difficult to use high frequency VCO (Voltage Controlled Oscillator) is averted. Moreover, the long-term stability and phase noise characteristics achievable with DDS chips are difficult to attain using a VCO. See [45 App. A] and [46] for an explanation on the DDS's operation.

The advantages of implementing the phase detector and the loop filter digitally are numerous. First and perhaps foremost, the repeatability of filter and transient response specifications that can be achieved via a digital implementation is exceedingly difficult

-15-

to duplicate in an analog implementation. Second, arbitrarily complicated phase detector and filter structures may be implemented, whereby that complexity is only limited by the amount of logic and computing power available for the digital section's implementation. Finally in this very non-exhaustive list, the implementation of certain synchronization loop elements by digital means allows testability and probing with accuracy and availability that is hard to attain in a completely analog system; for example, if the loop filter is implemented digitally then its (digital) input may be monitored by a computer console or analyzed in real time using the FFT transform.

Due to the low sampling rate requirements and the relatively low complexity in both the analog and digital domains, the hybrid implementation is the architecture of choice for high-datarate communications. Thus, it is of most interest to us. Hence, as already stated, we shall assume for the remainder of the book that the receiver has a hybrid architecture, whose precise characteristics are discussed in the next section.

1.4   Signal and Receiver Models that are Used in this Book

In this section we shall present the M-PSK receiver model that is discussed in the remainder of the book. In this book we are not concerned with implementation of the symbol synchronization PLL, and, unless otherwise stated, that PLL is assumed to be operating ideally. We thus limit ourselves to investigation of the carrier PLL, which is assumed to be a hybrid PLL as discussed in Sec. 1.3.3. Though a hybrid structure is assumed, it is commented that the structures presented in this book will work and will have the exact same behaviour if the carrier PLL has a digital topology. The simplified receiver model is as shown in Fig. 13, and this model is used throughout the book.

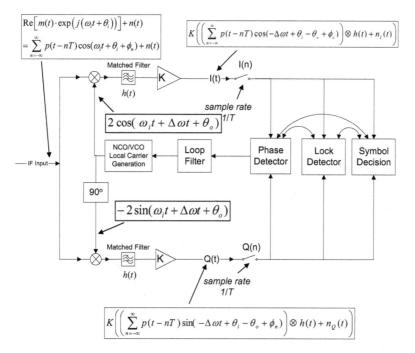

Fig. 13. M-PSK receiver model that is used in this book.

We denote the M-PSK data signal as $m(t) = \sum_{n=-\infty}^{\infty} a_n p(t-nT)$, where:

$$a_n = \exp\left(j\phi_n\right),\ \phi_n = 2\pi \cdot m_n / M,\ m_n \in \{0,1,...,M-1\} \tag{1}$$

is the actual data and $p(t)$ is the baseband data pulse. The modulated signal is

$s_m(t) = \text{Re}[m(t)e^{j\omega_i t + j\theta_i}]$ and that signal is corrupted by an AWGN channel. Fig. 13 shows a simplified diagram of the M-PSK receiver under discussion. In Fig. 13:

1. $1/T$ is both the symbol rate and the sample rate.

2. We assume a narrowband bandpass signal, i.e. $\omega_i \gg 1/T$.

3. $n(t) \sim N\left(0, N_0 W\right)$ where W is the width of the bandpass IF filter (not shown).

4. K represents the physical gain associated with the circuit. It is assumed that K has the same value in both the I and Q arms (i.e. the arms are "balanced"). In general, K is a slow function of time controlled by the AGC to achieve a desired signal level at the sampler inputs. A more detailed discussion of the AGC and the parameter K is presented in Section 1.5.

5. When the carrier loop is locked around a stable equilibrium point, we have $\Delta\omega = 0$ and (since M-PSK carrier synchronization has an inherent M-fold phase ambiguity ([13 Chap. 6], [22 Chap. 5, 6], [21 Sec. 5.7])) $\theta_o \in \{\theta_i + 2\pi k / M - \theta_e | k = 0,1,...,M-1\}$, where $|\theta_e| < \pi / M$ is the residual phase error.

6. The matched filter $h(t)$ is assumed ideal and the sampling at the outputs of matched filters is considered to be at the ideal time (i.e. the symbol synchronization loop is assumed locked).

7. From [13 Sec. 4.1.1] we have $E_S = \int_{-\infty}^{\infty} \left[p(t)\cos(\omega_i t + \theta_i) \right]^2 dt \approx \frac{1}{2} \int_{-\infty}^{\infty} p^2(t) dt = \frac{1}{2} E_p$.

   Without loss of generality, we assume for convenience $E_P = 1$ (implying $E_S = \frac{1}{2}$).

8. Throughout this book, the terms SNR and E_S / N_0 ratio are used interchangeably, and we use the notation χ to refer to the E_S / N_0 ratio (=SNR).

9. We assume that the Nyquist criterion for zero-ISI [13 Sec. 9.2.1] is obeyed regarding the output of the matched filters. Two important pulse shapes that fulfill this condition are:

• The rectangular pulse:

$$p(t) = \begin{cases} \sqrt{1/T} & -T/2 \leq t \leq T/2 \\ 0 & \text{otherwise} \end{cases} \tag{2}$$

• The Square-Root Raised Cosine (SRRC) pulse ([47 eq. 68.15]) which is shown in Fig. 14 (where $0 < \alpha \leq 1$ is the rolloff factor) :

$$p(t) = 4\alpha \frac{\cos\big((1+\alpha)\pi\, t/T\big) + \dfrac{\sin\big((1-\alpha)\pi\, t/T\big)}{4\alpha\, t/T}}{\pi\sqrt{T}\,\big(1 - 16\alpha^2\, t^2/T^2\big)} \tag{3}$$

10. Throughout this book we make the standard assumption made in synchronization texts (e.g. [22], [21]) that the carrier PLL is a high-loop-gain second-order system. Hence, the linearized-model Laplace transfer function of the PLL is

   $H(s) \triangleq \theta_o(s)/\theta_i(s) = \dfrac{2\zeta\omega_n \cdot s + \omega_n^2}{s^2 + 2\zeta\omega_n \cdot s + \omega_n^2}$ where ζ is the damping ratio and ω_n is the

   natural frequency in radians/sec (see for example [25 Sec. 2.2]).

11. We assume, unless otherwise stated, that no signal fading is occurring. Hence, the E_S/N_0 ratio is considered to be a constant. Fading effects shall be treated in a specific manner where appropriate.

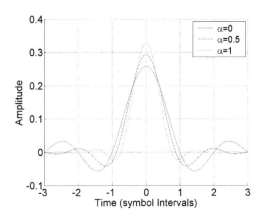

Fig. 14. Square-Root Raise Cosine pulse shapes

Note that in Fig. 13 we have:

$$n_I(nT) = \Big[\big(2n(t)\cos(\omega_i t + \Delta\omega t + \theta_o)\big) \otimes h(t)\Big]_{t=nT} \sim N(0, 2N_0 E_S)$$
$$n_Q(nT) = \Big[\big(-2n(t)\sin(\omega_i t + \Delta\omega t + \theta_o)\big) \otimes h(t)\Big]_{t=nT} \sim N(0, 2N_0 E_S) \tag{4}$$

and:

$$I(n) = K\left(2E_S \cos\left(-\Delta\omega \cdot nT + \theta_i - \theta_0 + \phi_n\right) + n_I(nT)\right)$$
$$Q(n) = K\left(2E_S \sin\left(-\Delta\omega \cdot nT + \theta_i - \theta_0 + \phi_n\right) + n_Q(nT)\right)$$

$$(5)$$

We shall now derive (5). Referring to the I channel, we have from Fig. 13:

$$I(t) = K\left(\left(\sum_{r=-\infty}^{\infty} p(t-rT)\cos(-\Delta\omega t + \theta_i - \theta_o + \phi_r)\right) \otimes h(t) + n_I(t)\right)$$

$$= K\left(\int_{-\infty}^{\infty}\left(\sum_{k=-\infty}^{\infty} p(\tau-rT)\cos(-\Delta\omega\tau + \theta_i - \theta_o + \phi_r)h(t-\tau)\right)d\tau + n_I(t)\right) \quad (6)$$

$$= K\sum_{r=-\infty}^{\infty}\int_{-\infty}^{\infty}\left(\cos(-\Delta\omega\tau + \theta_i - \theta_o + \phi_r)p(\tau-rT)h(t-\tau)\right)\cdot d\tau + Kn_I(t)$$

We shall limit ourselves to dealing with baseband pulses $p(t)$ that are real. Furthermore, we require that $p(t)$ have a finite effective length defined as $L = mT$, where m is some (small) positive integer, chosen so that $p(t)$ is nonzero only in the interval $[-L/2, L/2]$. This condition on $p(t)$ does not limit the applicability of the ensuing analysis, since the vast majority of practical systems employ baseband pulses which comply with this requirement or are closely approximated for the purposes of analysis by such finite duration pulses (this is true, for example, for both rectangular and SRRC pulses, given in (2) and (3), respectively). Under those conditions we have that the matched filter $h(t)$ abides by $h(t) = p(-t)$. Furthermore, for convenience we define $s(t) = p(t) \otimes h(t)$, and we note that we have assumed that $s(t)$ conforms to the Nyquist criterion for zero-ISI [13 Sec. 9.2.1]. From [13 Chap. 9] we thus have for any integer r:

$$s(rT) = \begin{cases} E_P(=2E_S) & \text{for } r = 0 \\ 0 & \text{for } r \neq 0 \end{cases}$$

$$(7)$$

For the ensuing analysis to be valid, we must assume that the beat note $\Delta\omega$ has a much smaller frequency than the matched filter bandwidth. This is not a real limitation since if this condition is not obeyed, then the I and Q signals cannot even be considered baseband signals and a coarse (open-loop) correction of the local carrier is needed if there is to be any hope of the loop acquiring lock. Formulation of this condition takes the following form:

$$\frac{|\Delta\omega|}{2\pi} << \frac{1}{L} = \frac{1}{mT} \tag{8}$$

Re-examining (6), we note that since $h(t)$ is nonzero only in the interval $[-L/2, L/2]$, the integration bounds can be reduced to comprise only the interval $[t - L/2, t + L/2]$. Furthermore, because of (8), the signal $\cos(-\Delta\omega\tau + \theta_i - \theta_o + \phi_r)$ can be assumed constant in the interval $[t - L/2, t + L/2]$ with the approximate value of $\cos(-\Delta\omega t + \theta_i - \theta_o + \phi_r)$. Using these observations, we can simplify (6) to:

$$I(t) = K \sum_{r=-\infty}^{\infty} \left(\cos(-\Delta\omega t + \theta_i - \theta_o + \phi_r) \int_{t-L/2}^{t+L/2} p(\tau - rT)h(t - \tau)d\tau \right) + Kn_I(t)$$

$$= K \sum_{r=-\infty}^{\infty} \cos(-\Delta\omega t + \theta_i - \theta_o + \phi_r)s(t - rT) + Kn_I(t) \tag{9}$$

The n-th symbol is sampled at time $t = nT$, so that:

$$I(n) = I(t)\big|_{t=nT} = K \sum_{r=-\infty}^{\infty} \cos(-\Delta\omega \cdot nT + \theta_i - \theta_o + \phi_r)s(nT - rT) + Kn_I(nT) \tag{10}$$

Because of (7), only one term in the summation does not vanish (the one for which $k = n$), and we are left with:

$$I(n) = K \left(\cos(-\Delta\omega \cdot nT + \theta_i - \theta_o + \phi_n)s(0) + n_I(nT) \right)$$

$$= K \left(2E_S \cos(-\Delta\omega t \cdot nT + \theta_i - \theta_o + \phi_n) + n_I(nT) \right) \tag{11}$$

which agrees with the I component in (5). A similar analysis can be carried out to prove the equation for the Q component in (5).

1.5   The AGC's Operation

General analysis of AGC-induced effects is hindered by the fact that the constraints and parameters of AGC circuits are strongly dependent upon the specific communications system. It is thus, perhaps, less of a surprise that most contemporary synchronization texts ignore these effects (by assuming a constant $K=1$). This is the case, for example, throughout [22] and [21], which are some of the most comprehensive modern works on synchronization in wireless communications. Nonetheless, some treatment of AGC effects does exist; see for example [48 Chap. 9] and [49 Chap. 7], though the discussions there pertain to unmodulated carrier-wave synchronization

For this book we wish to attain an understanding of the AGC's operation and effects upon the carrier PLL and associated structures. To that end, it is instructive to take a look at the waveforms before and after the AGC. We assume throughout this book an example AGC that attempts to control the waveform amplitudes at the input of the samplers so that the RMS value of the waveforms is 80% of the dynamic range of the samplers. We also assume, for simplicity, that the samplers' full-scale input range is ±1 volt, which means that our AGC tries to ensure that the pre-sampler waveforms have an RMS of 0.8 volt. The characteristics of this example AGC shall be discussed further in the following subsections.

1.5.1   The parameter K and its relationship with the AGC

Obtaining a physical insight into the meaning of the parameter K (see Fig. 13) is straightforward and should perhaps even be intuitively apparent to persons who have designed and built an M-PSK wireless receiver. To explicitly spell out this meaning, we recall that we assumed in Sec. 1.4 that for convenience and without loss of generality the baseband pulse energy (=matched filter energy) is unity, i.e. that $E_p = 2E_S = \int_{-\infty}^{\infty} p^2(t)dt = \int_{-\infty}^{\infty} h^2(t)dt = 1$. Let us further assume that we are ideally locked (i.e. $\Delta\omega = 0$ and $\theta_o \in \{\theta_i + 2\pi k / M \mid k = 0,1,..., M - 1\}$). We then have from (5) that:

$$I(n) = K\left(\cos\phi_n + n_I(nT)\right) \text{ and } Q(n) = K\left(\sin\phi_n + n_Q(nT)\right). \tag{12}$$

Now, define the time-average operator $\langle\bullet\rangle$ as $\langle x(n)\rangle \triangleq \lim_{N\to\infty} \frac{1}{2N}\sum_{n=-N+1}^{N} x(n)$. It is then easy to see that:

$$K \xrightarrow{E_S/N_0 \to \infty} \sqrt{\langle I^2(n) + Q^2(n)\rangle} \tag{13}$$

i.e. at high SNR we have that K is roughly the RMS (root-mean-square) of the M-PSK signal. Now, to inject a little more real-world issues into the model, we know that samplers have a finite number of bits and the AGC's job is, as already noted, to ensure that the samplers are not overdriven or underdriven. Consider, for example, a system which samples the input I and Q channels with 8-bit samplers, which give a range[1] for $I(n)$ and $Q(n)$ of ±127. Let us also assume that the AGC controls the input signal so that, to avoid sporadic overdriving the samplers due to noise, the input signal's RMS is controlled to about 80% of the dynamic range. Note that 8-bit samplers and 80% driving of the samplers are certainly real-world parameters. In terms of the signal model used in the book (see Fig. 13) where unity gain samplers are assumed, it can be seen that at high SNR the model of Fig. 13 applies with about $K=100$ (we rounded to $K=100$; the exact number is $K=80\% \times 127 = 101.6$).

Let us look at the situation at low SNR. Since we assumed for our model a constant E_S, it follows that $\text{var}(n_I(nT)) \xrightarrow{E_S/N_0 \to 0} \infty$ and $\text{var}(n_Q(nT)) \xrightarrow{E_S/N_0 \to 0} \infty$. The AGC still needs to control K so that the samplers are not overdriven. Hence, at low SNR, to insure finite and non-overdriving sampler inputs, we must have $K \xrightarrow{E_S/N_0 \to 0} 0$.

In summary then, if b is the number of bits in the samplers (including sign bit) and the AGC attempts to control the RMS of the input signal to $100 \cdot r$ percent of the samplers' range, we have $K \xrightarrow{E_S/N_0 \to 0} 0$ and $K \xrightarrow{E_S/N_0 \to \infty} r \cdot 2^{b-1}$. For example, for the 8-bits samplers with 80% driving we discussed above, we have $b=8$ and $r=0.8$, and the dynamic range of K is about $0 \leq K \leq 100$.

We could have just as easily adopted the convention that the binary point at the output of the samplers immediately follows the sign bit. For example, if the output of the sampler is 00000011_b, this could signify the value 3 or, alternatively, $\frac{3}{128}$ (if we think of the binary point as being at the right of the sign bit, i.e. 0.0000011_b). *This decision is purely arbitrary and has no actual bearing upon the resultant dynamic range analysis stemming from K's behaviour and the influence of the latter upon the rest of the receiver.* Under the convenient assumption of the binary point being after the sign bit of the sampler, we have that $0 \leq K \leq r \leq 1$ regardless of the number of bits in the sampler.

---

[1] More accurately, in a two's complement 8-bit system, we would have $I(n)$, $Q(n) \in \{-128...127\}$. For simplicity and fluency of the discussion (and without incurring any appreciable loss of accuracy), we ignore the −128 value.

Unless otherwise stated, this is the convenient implicit notational choice made in this book.

1.5.2   Demonstration of the AGC's operation

For the purposes of the following demonstration we make the convenient assumption that we are ideally locked (i.e. $\Delta\omega = 0$ and $\theta_o \in \left\{\theta_i + 2\pi k / M \middle| k = 0,1,...,M-1\right\}$). Now, let us look at the pre-AGC (=post-matched-filter) and the post-AGC (=pre-sampler) waveforms at various SNRs. We shall look at the I-channel of a BPSK signal that has rectangular baseband pulses (remember that the signal has passed through a matched filter, so the signal waveform is now composed of the triangular pulses $p(t) \otimes h(t)$). The sampling instances in the following figures are denoted by small circles.

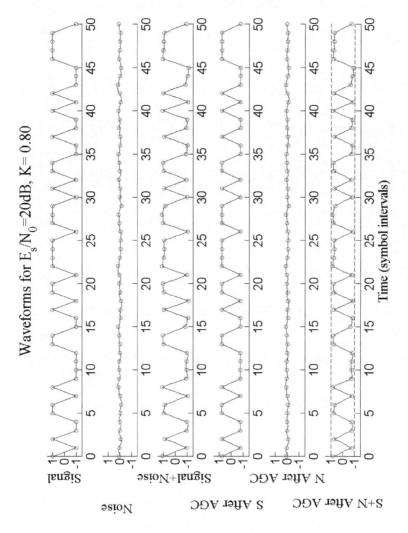

Fig. 15. Pre-AGC and post-AGC sampled waveforms for $E_S/N_0 = 20$ dB.

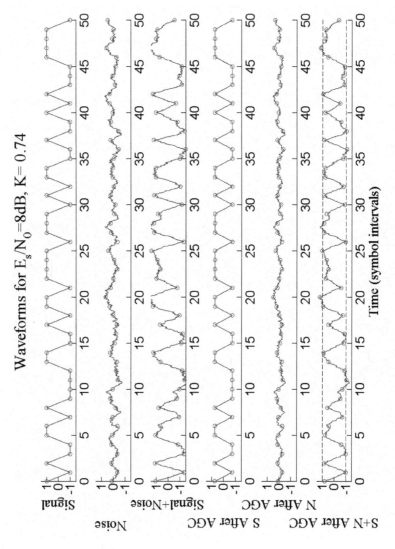

Fig. 16. Pre-AGC and post-AGC sampled waveforms for $E_s/N_0 = 8$ dB .

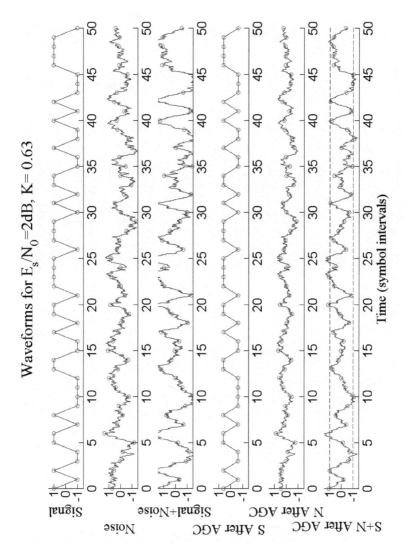

Fig. 17. Pre-AGC and post-AGC sampled waveforms for $E_S/N_0 = 2$ dB.

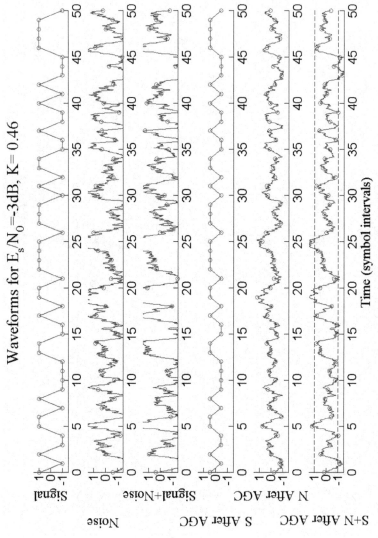

Fig. 18. Pre-AGC and post-AGC sampled waveforms for $E_S/N_0 = -3$ dB.

In the above graphs, we see the effects of the AGC upon the waveforms. In the bottom subplots, the dashed horizontal lines represent the samplers' full-scale voltage. Of course, the AGC "sees" only the pre-AGC signal+noise waveform, and the samplers "see" only the post-AGC signal+noise waveform; the separation into separate signal and noise waveforms that is shown in the above figures is only presented, courtesy of the computer simulation, for the benefit of the reader.

As can be clearly seen in the graphs, the AGC insures that the signal+noise waveform at the input of the samplers is such that the samplers are saturated infrequently. At $E_S / N_0 = -3\ dB$, which is just about the carrier PLL lock threshold for BPSK, we see from inspection of Fig. 18 that the samplers will be saturated only 10-12 samples out of 100 samples, i.e. only at most about 12% of the time, which may have an acceptably small effect on the receiver (results for higher SNRs are, of course, even better, as Fig. 15-Fig. 17 show). Yet, the 80% RMS driving level ensures that most of the dynamic range of the samplers is used at all SNRs (hence minimizing the quantization noise). Clearly, had K not been reduced to accommodate for the noise power, the samplers would be saturated almost all the time, especially at low SNRs, as can be seen in the signal+noise waveforms before the AGC for $E_S / N_0 = -3\ dB$ and $E_S / N_0 = 2\ dB$ (in Fig. 18 and Fig. 17, respectively).

Now, notice that the assumption made here is of a constant E_S and a changing noise power. Put another way, we assume that the E_S / N_0 changes because N_0 changes. This is contrary to what happens in practice. In practice, the noise power is generally constant (we have $NoisePowerPerHz_{dBm/Hz} = ThermalNoisePowerPerHz_{dBm/Hz} + ReceiverNoiseFigure_{dB}$) and the E_S / N_0 changes because E_S changes. However, the adoption of the convention of a constant E_S does not impact the analysis, and in fact simplifies it. The analysis is simplified because we can assume that a true matched filter (with energy $E_P = 2 \cdot E_S$) is present at the receiver. The receiver model adopted here is also the one used in most communications and synchronization texts (for example, see [22] and [21]). More importantly, the post-AGC (=pre-sampler) waveforms presented in Fig. 15 to Fig. 18 will be those that will indeed be encountered in practice.

Our AGC tries to control the RMS, namely it tries to control the value of $\sqrt{\left\langle I^2(n) + Q^2(n) \right\rangle}$, where $\left\langle \bullet \right\rangle$ is the time-average operator defined as $\left\langle x(n) \right\rangle \triangleq \lim_{N \to \infty} \dfrac{1}{2N} \sum_{n=-N+1}^{N} x(n)$. Therefore, this is an AGC that is a squaring detection AGC. Note that, despite appearances, this is not an Envelope Detection AGC. An Envelope Detection AGC would try to control the value of $\left\langle \sqrt{I^2(n) + Q^2(n)} \right\rangle$. In contrast, our AGC tries to control the value of $\sqrt{\left\langle I^2(n) + Q^2(n) \right\rangle}$ so that it equals a certain voltage v. This is equivalent to trying to control the value of $\left\langle I^2(n) + Q^2(n) \right\rangle$ so that it equals v^2. That having been said, the graphs for envelope detection and square-law detection are very similar, as [49 Fig. 7.2-5] shows, so the distinction is almost immaterial.

As further verification of the validity of the AGC behaviour presented here, we can define the *AGC signal suppression factor* at $E_S / N_0 = \chi$ as $\alpha_{AGC}(\chi) \triangleq \dfrac{K(E_S / N_0 = \chi)}{K(E_S / N_0 = \infty)}$.

The graph of $\alpha_{AGC}(\chi)$ for our example AGC is shown below.

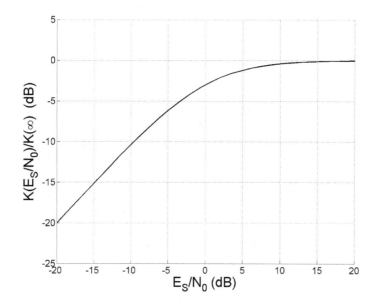

Fig. 19. AGC signal suppression factor for the example AGC used in this book.

Comparing Fig. 19 to [49 Fig. 7.2-5] we see that the AGC curve presented here is indeed in agreement with the data presented in [49] for the Squaring Detection AGC. The congruence between the curves is perfect and serves to highlight the validity of the AGC parameters used in this book.

1.5.3   Discussion – ideal AGC vs. atrophied AGC

It is of paramount importance to realize that we are still assuming an *ideal* AGC, i.e. the example AGC circuit discussed above is assumed to be devoid of lag time and is assumed to control the RMS of the pre-sampler waveforms to precisely 80% of the samplers' dynamic range. The assumption of a constant $K=1$, though undertaken in the vast majority of synchronization texts (e.g. [22], [21]) does not really describe an *ideal* AGC, but rather an *atrophied* AGC which operates within a system whose samplers have an infinite dynamic range and an infinite number of quantization bits. Clearly, the AGC discussed here (though still "ideal") is a much closer approximation of reality, as opposed to the assumption of a constant $K=1$.

Chapter 2   A Family of Self-Normalizing Carrier Lock Detectors and E_S/N_0 Estimators for M-PSK

2.1   Introduction

When building coherent M-PSK receivers, there is an invariable need to generate a reliable carrier lock detection mechanism. This is necessary, for example, in order for the receiver to know when to stop searching the carrier frequency uncertainty region in order to avoid driving the receiver out of lock. As another example, lock detection is necessary in order to allow the receiver to change (adaptively) the loop-filter parameters for different responses for acquisition and for tracking. Yet another use of lock detection is as a start trigger for downstream decoding and data processing elements of the receiver.

Generally, lock detection incurs generating a lock metric, which is compared to a threshold. If that threshold is exceeded, then lock is assumed; otherwise the receiver is considered to be unlocked. This process is illustrated in Fig. 20.

This chapter was published in part in Linn,Y. and Peleg, N., "A Family of Self-Normalizing Carrier Lock Detectors and E_S/N_0 Estimators for M-PSK and Other Phase Modulation Schemes", *IEEE Transactions on Wireless Communications*, vol. 3, no. 5, pp.1659-1668, Sept. 2004.

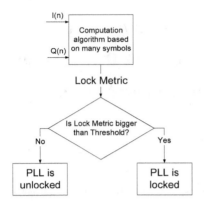

Fig. 20. Lock detection principle

Lock detectors such as those suggested in [50], [51], [48 Chap. 11], [22 Sec. 6.5.2], [25 Sec. 5.4], [52] and [53] operate according to the principle of Fig. 20. The prevalent methods for carrier lock detection for M-PSK rely either on Non Data Aided (NDA) detection based on M^{th}-order nonlinearities ([50], [51]) or via Decision Directed (DD) schemes [22 Sec. 6.5.2]. However, a drawback of the detectors in the aforementioned references is that the lock detector output is dependent upon the input signal level[2], and thus the threshold must be so dependent. This regularly overlooked problem often consumes a disproportionate amount of engineering time and energy during receiver design, in order to accommodate the dynamic range of the lock detector and to avoid false locking or false-unlocking due to non-ideal signal levels or non-ideal AGC performance. Even with an ideal AGC circuit any change in the AGC's nominal level must evoke a corresponding change in the lock detection threshold. In contrast, the lock detectors suggested in this chapter are self-normalizing, that is the lock threshold's value may be set independent of signal level or AGC performance.

---

[2] The term signal level as it is used in this book must not be confused with the term E_s / N_0 ratio. The former refers to the total signal+noise power that is present at the inputs of the samplers in Fig. 13, while the latter refers to the signal-to-noise ratio of that signal.

As a useful by-product, the value of the proposed lock metrics, when in lock, will be shown to be a reliable indicator of the E_S / N_0 ratio, which is another important metric in receiver operation. Frequently, a downstream decoder can make use of an E_S / N_0 estimate for modifying its own internal metrics in order increase its coding gain (e.g. turbo codes [12], diversity reception [13 Sec. 14.4]). Alternatively, the use of adaptive coding schemes (e.g., [14], [15]), where the coding and/or data rate is altered according to the channel E_S / N_0, presupposes the availability of a reliable and timely E_S / N_0 estimate. What is appealing about the proposed metrics is that they provide just such an estimate based only on the sampled baseband input signal (sampled at a rate of one sample per symbol, and that sample corresponds to the symbol strobe), necessitate a relatively small number of samples in order to arrive at an accurate estimate, and are irrespective of the data sequence. This obviates the need to estimate the channel E_S / N_0 from the pre- or post-decoder symbol or bit error rate of the received sequence, as is often done.

Finally, the lock detectors presented here will also be shown to have an extremely simple hardware implementation that requires only a single, compact lookup table and use of summation as the only mathematical operation, thus greatly facilitating implementation as part of an FPGA-based or ASIC-based receiver.

With regards to the layout of this chapter, it is as follows. Section 2.2 presents an overview of the general structure of the receiver and signal around which the discussion applies. Section 2.3 engages in rigorous statistical characterization of the lock metric in an AWGN channel, assuming perfect (i.e. jitter-free) carrier and symbol synchronization. That section further outlines the lock detector's hardware implementation and discusses E_S / N_0 estimation from the lock metric value, and it culminates in a subsection that aims to provide the reader with an intuitive insight into the lock metric's behaviour. Section 2.4 analyzes the lock detector's performance when imperfect locking is present, providing a detailed quantitative treatment for the case of carrier synchronization phase jitter. In Section 2.5 the variance and distribution of the lock metric are investigated. Section 2.6 provides design formulas for determining the lock detector parameters that will result in desired lock detection probabilities and false alarm rates. Section 2.7

-35-

discusses the lock detector's operation in the presence of fading. The final section in this chapter, Section 2.8, is devoted to conclusions.

2.2   Signal and Receiver Models

The signal and receiver models, as well as the applicable notations, have been defined in Section 1.4.

When in this chapter the terms "signal-level dependence" are referenced, the meaning pertains to dependence on K. Since K multiplies both the signal and noise, it is clear that any dependence of the lock detector characteristics or threshold on K is a mathematical appendage, as well as a practical one, because it introduces K (and its dynamic range) as a quantity to be reckoned with during the lock detection process. Furthermore, since K is a function of time controlled by the AGC (see Sec. 1.5), any dependence on K implies dependence of the lock detection process on the AGC loop's temporal behaviour, often through a decidedly nonlinear interaction. And yet, with previously available lock detectors such as those presented in [50], [51], [48 Chap. 11], [22 Sec. 6.5.2], [25 Sec. 5.4], and [53], precisely that kind of dependence exists. In contrast, the lock detectors suggested here are (as shall be shown shortly) absent of any association with K, and hence they and their thresholds are nearly impervious to the AGC's performance or dynamic range.

2.3   Detector Characteristics

2.3.1   Basic Definitions and Equations

The family of lock detectors is defined as:

$$\hat{l}_{M,N} \triangleq \frac{1}{2N} \sum_{n=-N+1}^{N} \frac{\operatorname{Re}[(I(n)+j \cdot Q(n))^{M}]}{\left(I^{2}(n)+Q^{2}(n)\right)^{\frac{M}{2}}} = \frac{1}{2N} \sum_{n=-N+1}^{N} \frac{\sum_{k=0}^{M/2}\binom{M}{2k}(-1)^{k} I^{M-2k}(n)Q^{2k}(n)}{\left(I^{2}(n)+Q^{2}(n)\right)^{\frac{M}{2}}} \quad (14)$$

which is of course a finite approximation of:

$$\hat{i}_{M,\infty} \overset{\triangle}{=} \lim_{N \to \infty} \hat{i}_{M,N} = \left\langle \frac{\mathrm{Re}\left[(I(n)+j\cdot Q(n))^M\right]}{\left(I^2(n)+Q^2(n)\right)^{\frac{M}{2}}} \right\rangle \tag{15}$$

where $\langle \bullet \rangle$ represents the time average operator defined as:

$$\langle x(n) \rangle \overset{\triangle}{=} \lim_{N \to \infty} \frac{1}{2N} \sum_{n=-N+1}^{N} x(n) \tag{16}$$

For example, for $M=4$ (QPSK) we have $\hat{i}_{4,N} \overset{\triangle}{=} \dfrac{1}{2N} \displaystyle\sum_{n=-N+1}^{N} \dfrac{I^4(n)-6I^2(n)Q^2(n)+Q^4(n)}{\left(I^2(n)+Q^2(n)\right)^2}$.

We shall also define for future convenience:

$$x_{M,n} \overset{\triangle}{=} \mathrm{Re}\left[(I(n)+j\cdot Q(n))^M\right] \Big/ \left(I^2(n)+Q^2(n)\right)^{\frac{M}{2}} \tag{17}$$

whereupon we have:

$$\hat{i}_{M,N} = \frac{1}{2N} \sum_{n=-N+1}^{N} x_{M,n} \tag{18}$$

The proposed lock detector can be thought of as a modification of the Mth- order nonlinearity detector ([50], [51], [22 Sec. 6.5.2]). To see this, consider that if the denominator term in (14) is eliminated, the latter equation reduces to $\dfrac{1}{2N} \displaystyle\sum_{n=-N+1}^{N} \mathrm{Re}[(I(n)+j\cdot Q(n))^M]$, which is the Mth- order nonlinearity detector. Generally, one can say that the denominator term in (14) performs adaptive normalization on the numerator; this action has a profound influence on the lock detector's statistics and implementation and makes it, despite the notational similarity, quite different from the Mth- order nonlinearity detector. These performance and implementational advantages shall be investigated in the remainder of this chapter.

2.3.2   Lock Detector Expectation

Elementary rectangular-to-polar manipulations and the use of De Moivre's theorem[3] [54 eq. 6.9] yields:

$$\text{Re}\left[\left(I(n)+j\cdot Q(n)\right)^M\right]=\left(\sqrt{I^2(n)+Q^2(n)}\right)^M\cdot\text{Re}\left[\left(\cos\varphi_n+j\cdot\sin\varphi_n\right)^M\right]$$
$$=\left(I^2(n)+Q^2(n)\right)^{\frac{M}{2}}\cdot\cos\left(M\varphi_n\right) \tag{19}$$

where:

$$\varphi_n\triangleq\tan^{-1}\left(\frac{Q(n)}{I(n)}\right) \tag{20}$$

and using (5) we can write:

$$\varphi_n\triangleq\tan^{-1}\left(\frac{Q(n)}{I(n)}\right)=\tan^{-1}\left(\frac{\sin(-\Delta\omega\cdot nT+\theta_i-\theta_0+\phi_n)+n_Q(nT)/(2E_S)}{\cos(-\Delta\omega\cdot nT+\theta_i-\theta_0+\phi_n)+n_I(nT)/(2E_S)}\right) \tag{21}$$

If the carrier loop is locked and assuming perfect coherent demodulation, we have $\Delta\omega=0$ and $\theta_o\in\left\{\theta_i+2\pi k/M\,|\,k=0,1,...,M-1\right\}$, and hence from (5):

$$I(n)=K\left(2E_S\cos\,\psi_n+n_I(nT)\right)$$
$$Q(n)=K\left(2E_S\sin\,\psi_n+n_Q(nT)\right) \tag{22}$$

where:

$$\psi_n\triangleq\phi_n+\theta_i-\theta_0=\phi_n-2\pi k/M=2\pi(m_n-k)/M \tag{23}$$

Thus, when locked we have from (21)-(23):

$$\varphi_n=\tan^{-1}\left(\frac{\sin(\psi_n)+n_Q(nT)/(2E_S)}{\cos(\psi_n)+n_I(nT)/(2E_S)}\right) \tag{24}$$

(that is, when locked φ_n is a noise-perturbed estimate of ψ_n). The rationale behind (14)-(15) now immediately becomes apparent:

---

[3] DeMoivre's theorem states that for any real x and y, $(x+j\cdot y)^M=(x^2+y^2)^{M/2}\exp(j\cdot M\theta)$ where $\theta\triangleq\tan^{-1}(y/x)$.

$$\hat{i}_{M,\infty} = \left\langle \frac{\text{Re}\left[\left(I(n)+j\cdot Q(n)\right)^M\right]}{\left(I^2(n)+Q^2(n)\right)^{\frac{M}{2}}} \right\rangle = \left\langle x_{M,n} \right\rangle = \left\langle \frac{\left(I^2(n)+Q^2(n)\right)^{\frac{M}{2}}\cos(M\varphi_n)}{\left(I^2(n)+Q^2(n)\right)^{\frac{M}{2}}} \right\rangle \qquad (25)$$

$$= \left\langle \cos(M\varphi_n) \right\rangle$$

If the carrier loop is unlocked, then the variables φ_n are samples of phases of a rotating sinusoid (that sinusoid is noise-corrupted and phase-modulated, but it is nonetheless rotating). Thus:

$$\hat{i}_{M,N} \xrightarrow{N\to\infty} \hat{i}_{M,\infty} = \left\langle x_{M,n} \right\rangle = \left\langle \cos\left(M\varphi_n\right) \right\rangle \overset{\text{unlocked}}{=} 0 \qquad (26)$$

Further justification of (26) will be given in Section 2.3.6. If the carrier loop is locked, then noting that:

$$\cos\left(M(\varphi_n - \psi_n)\right) = \cos\left(M\varphi_n\right)\cos\left(2\pi(m_n - k)\right) + \sin\left(M\varphi_n\right)\sin\left(2\pi(m_n - k)\right)$$
$$= \cos\left(M\varphi_n\right) \qquad (27)$$

and denoting $\Delta\phi_n \triangleq \varphi_n - \psi_n$, we have:

$$\hat{i}_{M,N} \xrightarrow{N\to\infty} \hat{i}_{M,\infty} = \left\langle \cos\left(M\varphi_n\right) \right\rangle \qquad (28)$$

$$= \left\langle \cos\left(M\Delta\phi_n\right) \right\rangle \xrightarrow[\text{loop is locked}]{E_S/N_0 \to \infty \text{ and}} \left\langle \cos\left(M\cdot 0\right) \right\rangle = 1$$

When locked, the departure of $\hat{i}_{M,\infty}$ from the value of 1 is dependent on the channel E_S/N_0, and for the finite-approximation $\hat{i}_{M,N}$ this also depends on how large N is. Thus it is clear that $\hat{i}_{M,N}$ when the carrier loop is locked provides an estimate of the channel E_S/N_0. In order to develop quantitative results regarding the dependence of $\hat{i}_{M,N}$ on the E_S/N_0 ratio, we note that when in lock at $E_S/N_0 = \chi$ the process $\Delta\phi_n$ has the distribution as the phase of a Rice random variable with a probability density function (pdf) for $|\Delta\phi| \le \pi$ of ([55 Sec. 4.5], [56], [13 Sec. 5.2.7], [57 Sec. 3.4]):

$$p_R\left(\Delta\phi \middle| \chi\right) \triangleq p\left(\Delta\phi_n = \Delta\phi \middle| E_S/N_0 = \chi\right)$$

$$= \frac{\exp(-\chi)}{2\pi} \times \left[1 + \sqrt{2\chi}\cos(\Delta\phi)\exp\left(\chi\cdot\cos^2(\Delta\phi)\right)\cdot \int_{-\infty}^{\cos(\Delta\phi)\sqrt{2\chi}} e^{-y^2/2}dy\right] \qquad (29)$$

Noting that the distributions of the variables $\Delta\phi_n$ are identical (there is no dependence of the pdf on the actual symbol transmitted), and that these variables are mutually independent, we have assuming ergodicity and using (27):

$$f_M(\chi) \triangleq E\left[\hat{i}_{M,N} \middle| E_S / N_0 = \chi\right] = \hat{i}_{M,\infty}\bigg|_{E_S/N_0=\chi} = \left\langle \cos\left(M\varphi_n\right)\big|_{E_S/N_0=\chi}\right\rangle$$
$$= \left\langle \cos\left(M\Delta\phi_n\right)\big|_{E_S/N_0=\chi}\right\rangle = E\left[\cos\left(M\Delta\phi_n\right)\middle| E_S / N_0 = \chi\right] \qquad (30)$$
$$= \int_{-\pi}^{\pi} \cos\left(M\Delta\phi\right) \cdot p_R\left(\Delta\phi \middle| \chi\right) \cdot d\Delta\phi$$

It is easy to compute (30) numerically, but due to the complicated nature of $p_R\left(\Delta\phi \middle| \chi\right)$, as expressed in (29), computation of (30) may seem to elude closed-form representations. However it is in fact possible to arrive at a closed-form formulas for $f_M(\chi)$. This is done in Appendix A. We arrive there at the following closed-form representation:

$$f_M(\chi) = \frac{\sqrt{\pi \cdot \chi}}{2} \cdot \exp\left(\frac{-\chi}{2}\right)\left[I_{\frac{M-1}{2}}\left(\frac{\chi}{2}\right) + I_{\frac{M+1}{2}}\left(\frac{\chi}{2}\right)\right] \qquad (31)$$

where $I_k\left(\bullet\right)$ is the k-th order modified Bessel function of the first kind [54 Chap. 24]. Moreover, (31) can be simplified even more, as shown in Appendix A. We arrive at the following formula, which is a <u>finite</u> sum composed of polynomials and exponentials:

$$f_M(\chi) = 1 + \frac{M}{2}\sum_{n=1}^{M/2}\frac{(-1)^n}{n!}\cdot\frac{(M/2+n-1)!}{(M/2-n)!\,\chi^n}$$
$$+ \exp\left(-\chi\right)\left[(-1)^{M/2+1}\sum_{n=1}^{M/2}\frac{1}{(n-1)!}\frac{(M/2+n-1)!}{(M/2-n)!\,\chi^n}\right] \qquad (32)$$

As examples, the results of (32) are shown for $M = 2$, 4 and 8 in Table 1.

Table 1. Closed-form expressions for lock detector expected value

M	$f_M(\chi) \triangleq E\left[\hat{i}_{M,N} \middle\| E_S / N_0 = \chi\right]$
2	$f_2(\chi) = \left(1 - \dfrac{1}{\chi}\right) + \dfrac{\exp(-\chi)}{\chi}$
4	$f_4(\chi) = \left(1 - \dfrac{4}{\chi} + \dfrac{6}{\chi^2}\right) + \left(-\dfrac{2}{\chi} - \dfrac{6}{\chi^2}\right)\exp(-\chi)$
8	$f_8(\chi) = \left(1 - \dfrac{16}{\chi} + \dfrac{120}{\chi^2} - \dfrac{480}{\chi^3} + \dfrac{840}{\chi^4}\right) + \left(-\dfrac{4}{\chi} - \dfrac{60}{\chi^2} - \dfrac{360}{\chi^3} - \dfrac{840}{\chi^4}\right)\exp(-\chi)$

While (31), (32), and Table 1 are extremely useful in computer simulations and generation of lookup tables for estimation of the E_S / N_0 from $\hat{i}_{M,N}$, they somewhat tedious for use by the designer during the initial design process, e.g. with a handheld calculator in order to get a "rough draft" idea of the detector's behaviour. Thus a closed-form simplification is sought for this purpose. To that end, if we look at high E_S / N_0 ratios, we have that $\Delta\phi_n$ is generally very small (i.e. $p_R(\Delta\phi|\chi)$ is non-negligible only for small $\Delta\phi$), and thus we can write (using (29) and $\int_0^\infty e^{-\alpha x^2} dx = \frac{1}{2}\sqrt{\frac{\pi}{a}}$ [54 eq. 15.72]):

$$p_R(\Delta\phi|\chi) \approx \frac{1}{2\pi}\exp(-\chi) \times \sqrt{2\chi}\,\exp\left(\chi \cdot \cos^2(\Delta\phi)\right) \cdot \int_{-\infty}^{\infty} e^{-y^2/2}\,dy$$

$$\approx \sqrt{\frac{\chi}{\pi}}\,\exp\left(-\chi(\Delta\phi)^2\right) \qquad (33)$$

Which means that for high E_S / N_0 ratios:

$$\Delta\phi_n \sim N\left(0, \frac{1}{2\chi}\right) \tag{34}$$

(see also [56] for a similar derivation) from which it follows that (using

$$\int_0^\infty e^{-ax^2}\cos(bx)dx = \frac{1}{2}\sqrt{\frac{\pi}{a}}e^{-b^2/(4a)} \quad \text{[54 eq. 15.73]):}$$

$$f_M(\chi) = \int_{-\pi}^{\pi}\cos(M\Delta\phi)\cdot p_R\left(\Delta\phi\big|\chi\right)\cdot d\Delta\phi$$
$$\approx \sqrt{\frac{\chi}{\pi}}\int_{-\infty}^{\infty}\cos(M\Delta\phi)\cdot\exp\left(-\chi(\Delta\phi)^2\right)d\Delta\phi = \exp\left(\frac{-M^2}{4\chi}\right) \tag{35}$$

Fig. 21 shows the value of $\hat{l}_{M,8192}$ (obtained through simulation) and Eqs. (30) and (35) as a function of the E_S/N_0. As seen in that figure, (35) is a good approximation even for low E_S/N_0 ratios, which makes it quite a useful tool for the designer. Note also that the linear range of the curves corresponds precisely to the most "interesting" range of E_S/N_0 ratios, i.e. from the minimal ratio where lock can be maintained [58 Sec. III] to a moderately high E_S/N_0 for the modulation in question.

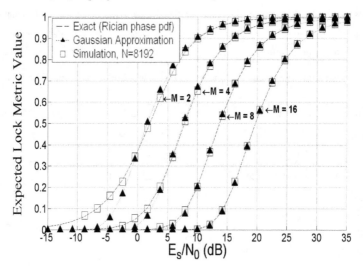

Fig. 21. Lock metric when locked vs. E_S/N_0.

2.3.3 Hardware Realization

$\hat{I}_{M,N}$ lends itself to efficient hardware implementation. This can be easily seen by looking at (14), (17), and (18): the terms $x_{M,n}$ can be generated by a single, fixed-point lookup table which has $I(n)$ and $Q(n)$ as its address, and a single digital Integrate-and-Dump module can perform the summation (the division by $2N$ is avoided if $2N$ is chosen to be a power of 2, and then the division can be implemented by discarding the lower $\log_2 2N$ bits of the output of the summation). Thus no mathematical operations except summation are required. This is depicted in Fig. 22.

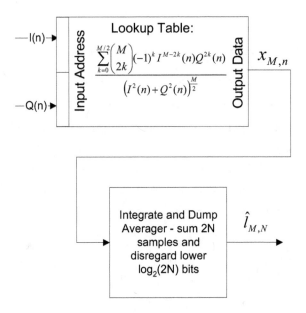

Fig. 22. Efficient hardware generation of $\hat{l}_{M,N}$.

Since[4] for all M, K and n, $\left|x_{M,n}\right| = \left|\cos\left(M\varphi_n\right)\right| < 1$, this implies that the lookup table in Fig. 22 needs only to facilitate representation of fractional numbers, hence making its implementation in hardware quite practical. Additionally, comments which are also easily applicable to the lookup table in Fig. 22 are given in Sec. 3.4.9 and are omitted here in order to avoid repetition. Those comments outline the practical way to implement the lookup table in fixed-point hardware, and the reader is referred to Sec. 3.4.9 for more information.

Now let us treat the other component of Fig. 22, namely the Integrate and Dump module. The Integrate-and-Dump (IAD) averager is one of the simplest modules available for hardware signal processing, and among the most widely used and well

---

[4] Ignoring the infinitesimally probable case of $\left|x_{M,n}\right| = 1$ which can always be approximated to any desired tolerance by suitably close fractions

known. It must be emphasized that the Integrate-and-Dump averager is <u>not</u> the same as a "moving average". This distinction is crucial, and is now explained.

First, let us look at a <u>moving average</u>. It is shown below:

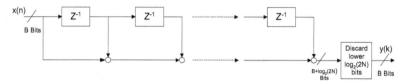

Fig. 23. Moving average: There are 2N-1 taps (registers). The current sample and 2N-1 delayed samples are summed at each clock; then (in a fixed-point hardware implementation) the $\log_2(2N)$ lower bits are discarded in order to produce the average $y(k)$.

Now, let us look at an <u>Integrate-and-Dump</u> module:

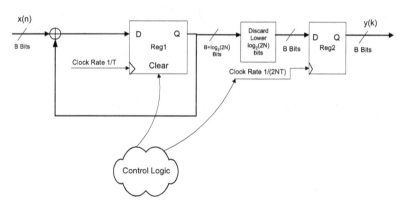

Fig. 24. Integrate and Dump module. There is one feedback register (Reg1) (the accumulator) which holds the accumulated value. Every 2N samples the value of Reg1 is passed on to the output $y(k)$ after the $\log_2(2N)$ lower bits are discarded. Then the value of Reg1 is cleared to make way for the next 2N samples.

Note that the moving average produces an output every T seconds, where $1/T$ is the rate of samples of $x(n)$. In other words, for the moving average the output clock rate of $y(k)$ is the same of that of $x(n)$. In contrast, the Integrate-and-Dump module produces an output sample every $2 \cdot N \cdot T$ seconds, i.e. the sampling rate of $y(k)$ is $\dfrac{1}{2 \cdot N}$ less than that of $x(n)$. This is the "price" we pay for the reduced complexity of the Integrate-and-Dump module (as opposed to the moving-average).

The use of an Integrate-and-Dump averager was deliberate. This is because a moving average would require inordinate amounts of logic resources. For example, consider the resources needed for the storage elements in the moving average for an 8-bit input data (i.e. x(n) is 8-bits wide) for N=1024. Using the moving average, we would need $8 \cdot (2N - 1) = 16376$ flip-flops and $2N - 1 = 2047$ adders - a truly remarkable amount of logic. However, using the Integrate-and-Dump module, we would need only two registers, Reg1 and Reg2, and only *a single adder*. Reg1 would have $8 + \log_2(2N) = 8 + 11 = 19$ flip-flops, and Reg2 would have 8 flip-flops. There would also be need for some "control logic" (see the Fig. 24). This logic is essentially little more than a self-resetting counter which controls the signals of the "clock" of Reg2 and the "clear" of Reg1. This control logic is, thus, trivial to implement. In summary, then, the Integrate-and-Dump averager implementation is very compact and offers a very significant logic savings as compared to the moving-average implementation.

2.3.4   Implementational Advantages vs. Other Lock Detectors

To see the advantage of the proposed detector over previously available detectors, we note that the M-PSK lock detectors which are prevalently used today are the following:

- The NDA (Non-Data Aided) M[th] power detector ([50], [51], [22 Sec. 6.5.2]):

$$LNDA_{M,N} \triangleq \frac{1}{2N} \sum_{n=-N+1}^{N} \mathrm{Re}[(I(n) + j \cdot Q(n))^M] \qquad (36)$$

- The DD (Decision Directed) detector [22 Sec. 6.5.2]:

$$LDD_{M,N} \triangleq \frac{1}{2N} \sum_{n=-N+1}^{N} \mathrm{Re}\left([\hat{I}(n) - j \cdot \hat{Q}(n)][I(n) + j \cdot Q(n)] \right)$$

$$= \frac{1}{2N} \sum_{n=-N+1}^{N} I(n)\hat{I}(n) + Q(n)\hat{Q}(n)$$

(37)

where $\hat{I}(n) + j \cdot \hat{Q}(n)$ is the symbol decision.

The fact that $\hat{l}_{M,N}$ can be computed using a small fixed-point lookup table is in sharp contrast to the lock detector schemes of (36) and (37), for reasons which we shall now outline. Dealing first with (36), we see that $LNDA_{M,N}$ value of the lock detector includes a dominant term that is proportional to K^M, and accordingly the lock threshold must be so dependent; additionally, any attempt to compute those lock detectors must accommodate the dynamic range of K^M, which quite often precludes their implementation through the use of fixed-point lookup tables in hardware. An example of the necessary implementation is shown in Fig. 25 for $LNDA_{2,N}$. Since K is a function of the AGC, $\hat{l}_{M,N}$'s independence from K also provides significant insulation from false locking and loss of lock due to non-ideal AGC behaviour, particularly when dealing with rapidly fading signals. Specifically, instead of the AGC having to cordon K to within the range $K_{nominal} - tolerance \le K \le K_{nominal} + tolerance$, the AGC now has only to abide by the relatively loose requirements that K is such that (a) no signal-chain or sampler saturation occurs, and (b) the samplers are not underdriven to the extent that quantization noise becomes significant.

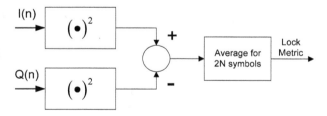

Fig. 25. 2nd-order nonlinearity lock metric generation for BPSK.

Dealing now with $LDD_{M,N}$ (given in (37)), we see that the latter is proportional to K. This may, depending on the receiver's implementation, present dynamic range problems which are similar (though not as acute) as compared to $LNDA_{M,N}$. Moreover, we observe that $LDD_{M,N}$'s value and threshold are dependent upon the AGC's nominal level and performance. In contrast, again, no such dependence exists for $\hat{l}_{M,N}$.

2.3.5 Principle of SNR Estimation from the Lock Detector Value

We can derive an approximation of the E_s / N_0 ratio from $\hat{l}_{M,N}$. Using units of dB we have from (35):

$$\chi_{dB} \approx 10 \cdot \log_{10}\left(f_M^{-1}\left(\hat{l}_{M,N} \right) \right) \approx 10 \log_{10}\left(\frac{-M^2}{4 \cdot \ln\left(\hat{l}_{M,N} \right)} \right) \tag{38}$$

If we indeed use units of dB, we have a greatly reduced dynamic range. Combined with the small dynamic range required to describe $\hat{l}_{M,N}$, this allows (38) to be implemented as a relatively small fixed-point lookup table, hence facilitating rapid, reliable estimation of E_s / N_0 within an FPGA or an ASIC. Note that if increased accuracy is required the right side of (38) can be replaced with the (numerically obtained) values derived from the lock metric expected value of (31)-(32). This can likewise be incorporated into a lookup table so no logic complexity penalty is incurred. Thus, estimation of the E_s / N_0 can be achieved using the method described here, using an almost trivial hardware structure, with one sample per symbol (which corresponds to the symbol strobe), and without the need for any symbol decisions to be made. This appears to be quite an improvement over previously available methods, as are analyzed in [59], [60], [61], [62], [63], [64], [65], [66], [67], [68], [69], [70], [71], [72], [73], [16] and [74]. The proposed SNR estimation method is investigated in depth in Chapter 4.

2.3.6   Intuitive Understanding of the Detectors' Behaviour

The behaviour of $\hat{l}_{M,N}$ can be intuitively understood by looking at a contour graph of

$x_{M,n}$. This is shown for the case of QPSK, i.e. $x_{4,n} = \dfrac{I^4(n) - 6I^2(n)Q^2(n) + Q^4(n)}{\left(I^2(n) + Q^2(n)\right)^2}$, in

Fig. 26 and Fig. 27, where sample demodulated signals are superimposed on contour maps of $x_{4,n}$.

Conforming to the title of this subsection, we shall now proceed with an intentionally non-mathematical discussion, with the hope of achieving an intuitive appreciation of the lock detector computation process. As exemplified in Fig. 26, for a given demodulated symbol $\left(I(n), Q(n)\right)$ the corresponding value of $x_{M,n}$ is an indication of how close the phase of that symbol is to a valid M-PSK constellation point's phase. If the carrier is locked (as is the case depicted in Fig. 26), the symbols $\left(I(n), Q(n)\right)$ will be concentrated in clouds around the constellation signal points, which, for QPSK, are on the I and Q axes (on the "1" contour). The location of the center of each cloud is a function of the signal level (i.e. is dependent upon K), however this does not impact the value of $x_{4,n}$ due to the radial symmetry of the contours. It is also intuitively clear that the size of the demodulated signal point clouds will be inversely related to the E_S / N_0. The ideal constellation points are on the "1" contour, and thus when the values are averaged for many symbols the value of the lock metric will be 1 for infinite E_S / N_0, and decrease as the E_S / N_0 decreases, as more symbols depart more significantly from the constellation point's phase and a bigger proportion of contours of lower values are encountered.

If the carrier is unlocked, the phase of the demodulated constellation will rotate due to the incommensurate local and received carriers. Consequently, the output $x_{M,n}$ of the lookup table will be concentrated on a circular ring around the origin, with the ring's radius (i.e. mean distance from the origin) proportional to K, and the ring's width inversely related to the E_S / N_0 ratio. This scenario is depicted in Fig. 27. The values accrued for many symbols will thus (from symmetry considerations) average to 0 in the unlocked state; this reasoning provides graphical validation to (26).

-49-

It is again emphasized that due to the radial symmetry that is evident from the contour graphs of Fig. 26 and Fig. 27, there is no dependence of $\hat{l}_{M,N}$ (in either the locked or unlocked case) on K but only on the angle of departure of the demodulated constellation from the ideal M-PSK constellation and on the E_S/N_0. To visually illustrate the difference between $\hat{l}_{M,N}$ and previously available lock detectors, refer to Fig. 28 and Fig. 29. In those figures demodulated QPSK signals, with $K = 0.8$ and $K = 0.4$ respectively, are superimposed upon a contour map of $\mathrm{Re}[(I(n) + j \cdot Q(n))^4]$, which is the NDA 4th power nonlinearity lock detector term (see [50], (36)). Whereas the radial symmetry evident in Fig. 26 ensured that variations in K did not have an impact on the value of $\hat{l}_{M,N}$, this is clearly not the case in Fig. 28 and Fig. 29.

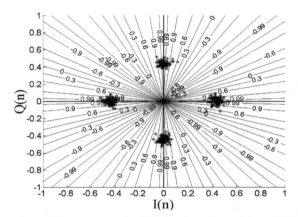

Fig. 26. Received QPSK signal, with $E_S/N_0 = 20dB$ and with receiver in lock, $K = 0.5$, superimposed on contour map of $x_{4,n}$.

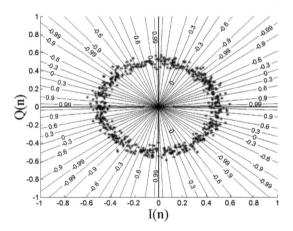

Fig. 27. Received QPSK signal, with $E_S/N_0 = 20dB$, unlocked receiver, $K = 0.5$, superimposed on contour map of $x_{4,n}$.

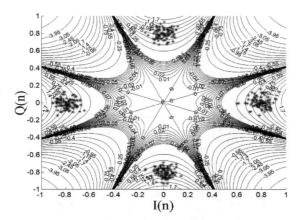

Fig. 28. Received QPSK signal, with $E_s / N_0 = 20dB$, $K = 0.8$, receiver in lock, superimposed on contour map of $\mathrm{Re}[(I(n) + j \cdot Q(n))^4]$.

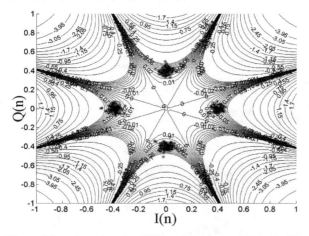

Fig. 29. Received QPSK signal, with $E_s / N_0 = 20dB$, $K = 0.4$, receiver in lock, superimposed on contour map of $\mathrm{Re}[(I(n) + j \cdot Q(n))^4]$.

2.4   Performance Analysis for Imperfect Locking

If the local carrier exhibits phase-jitter (i.e. imperfect carrier PLL lock), or the sampling of the outputs of the matched filters is not at the ideal instant (i.e. imperfect symbol PLL lock), this also degrades the value of $\hat{l}_{M,N}$, though this could also be modeled as an effective decrease of the E_S / N_0 ratio [19 Sec. 4.3.2].

The case of imperfect locking due to carrier phase jitter can also be easily modeled by modifying the definition of θ_o during lock to account for the residual phase error θ_e (where $|\theta_e| < \pi / M$). The modification is as follows:

$$\theta_o \in \left\{ \theta_i - \theta_e + 2\pi k / M \,\middle|\, k = 0,1,..., M-1 \right\} \tag{39}$$

With this definition we have when locked (using (21)):

$$\varphi_n = \tan^{-1} \left(\frac{\sin(\theta_e + 2\pi(m_n - k)/M) + n_Q(nT) / (2E_S)}{\cos(\theta_e + 2\pi(m_n - k)/M) + n_I(nT) / (2E_S)} \right) \tag{40}$$

Furthermore, we shall arbitrarily set $k = 0$. This is tantamount to ignoring the carrier synchronization ambiguity, or equivalently rotating the demodulated constellation by $2\pi k_0 / M$, where k_0 is an arbitrary integer. This is permissible due to the fact that the value of $x_{M,n}$ for a given demodulated symbol is invariant under such a rotation (see Section 2.3.6 and Fig. 26 for a graphical illustration of this property). Finally, as a generalization of (27), we note that for any M-PSK constellation point $\phi_x = 2\pi \cdot m_x / M$ where m_x is an integer we have:

$$\begin{aligned}
&\cos \left(M (\varphi_n - \phi_x) \right) \\
&= \cos \left(M \varphi_n \right) \cos \left(2\pi \cdot m_x \right) + \sin \left(M \varphi_n \right) \sin \left(2\pi \cdot m_x \right) \\
&= \cos \left(M \varphi_n \right) (= x_{M,n})
\end{aligned} \tag{41}$$

that is, the lock detector metric term for a given symbol is not dependent upon that symbol but rather only on the phase *difference* between the received symbol's phase φ_n and the phase of *any* valid M-PSK constellation point ϕ_x. Thus we can assume for simplicity $\phi_n = 0$ (which implies $m_n = \psi_n = 0$) for all n. Under the above simplifications we can rewrite (40) as:

$$\varphi_n = \tan^{-1}\left(\frac{\sin(\theta_e) + n_Q(nT)/(2E_S)}{\cos(\theta_e) + n_I(nT)/(2E_S)}\right) \tag{42}$$

A quick glance shows that (42) is of the same form as (24) with ψ_n replaced by θ_e. Therefore if we re-define:

$$\Delta\phi_n \triangleq \varphi_n - \theta_e \tag{43}$$

then $\Delta\phi_n$ is still distributed according to (29). Thus, if we define $\tilde{f}_M(\chi)$ to be the expected lock detector value at an SNR of $E_S / N_0 = \chi$ including carrier jitter effects, then:

$$
\begin{aligned}
\tilde{f}_M(\chi) &\triangleq E\left[\hat{i}_{M,N}\big| E_S/N_0 = \chi\right] = \hat{i}_{M,\infty}\big|_{E_S/N_0=\chi} = \left\langle x_{M,n}\big|_{E_S/N_0=\chi}\right\rangle \\
&= \left\langle \cos(M\varphi_n)\big|_{E_S/N_0=\chi}\right\rangle = E\left[\cos(M(\Delta\phi_n + \theta_e))\big|_{E_S/N_0=\chi}\right] \\
&= \left[E\left[\cos(M\Delta\phi_n)\right]\cdot E\left[\cos(M\theta_e)\right] - \underbrace{E\left[\sin(M\Delta\phi_n)\right]}_{=0}\cdot E\left[\sin(M\theta_e)\right]\right]_{E_S/N_0=\chi} \\
&= \left[E\left[\cos(M\Delta\phi_n)\right]\cdot E\left[\cos(M\theta_e)\right]\right]_{E_S/N_0=\chi} \\
&= f_M(\chi)E\left[\cos(M\theta_e)\big| E_S/N_0 = \chi\right]
\end{aligned}
\tag{44}
$$

where, through comparison with (30), we see that the presence of carrier phase jitter results in the degradation of the lock metric expected value by the multiplicative, less-than-unity factor of:

$$\upsilon(\chi) \triangleq E\left[\cos(M\theta_e)\big| E_S/N_0 = \chi\right] \tag{45}$$

The factor in (45) can be further simplified by using the Taylor expansion method [54 Chap. 20]. Using the Taylor expansion $\cos x = \sum_{n=0}^{\infty}(-1)^n \frac{x^{2n}}{(2n)!}$ [54 eq. 20.22] and adopting a first order approximation which retains only the first two terms of the Taylor series, we get:

$$\upsilon(\chi) = E\Big[\cos\big(M\theta_e\big)\big| E_S / N_0 = \chi\Big] = E\left[\sum_{n=0}^{\infty}(-1)^n \frac{(M\theta_e)^{2n}}{(2n)!}\bigg| E_S / N_0 = \chi\right]$$

$$\approx 1 - \frac{M^2 E\Big[\theta_e^2 \big| E_S / N_0 = \chi\Big]}{2}$$

(46)

and we can write:

$$\tilde{f}_M(\chi) = f_M(\chi)\upsilon(\chi) \approx f_M(\chi)\left(1 - \frac{M^2 E\Big[\theta_e^2 \big| E_S / N_0 = \chi\Big]}{2}\right)$$

(47)

and the phase error variance $E\Big[\theta_e^2 \big| E_S / N_0 = \chi\Big]$ can be evaluated using a plethora of known methods, such as those presented in [26], [49], [19], [51], [9], and [58].

To validate the results of (44)-(47), computer simulations of an equivalent (w.r.t. Fig. 13) baseband system were conducted, with the synchronization loop configured to behave as a second-order PLL. This model is shown in Fig. 30. In those simulations, the values of the lock detector were recorded, as was done with the values of the residual phase error θ_e. The recorded lock detector values were then compared to those computed using estimates of (44) and (47). The simulation results are shown in Fig. 31, Fig. 32, and Fig. 33, for $2B_L \cdot T = 0.01$, 0.1, and 0.25, respectively. B_L is the noise bandwidth of the carrier synchronization PLL (see [25 Sec. 3.1 and App. A] for a discussion of this parameter). For the second-order loop under discussion, $B_L = 0.5\omega_n\big(\zeta + 1/(4\zeta)\big)$ where ω_n is the natural radian frequency of the PLL and ζ is its damping factor [25 Sec. 3.1 and App. A]). As can be seen in Fig. 31- Fig. 33, is excellent agreement between (44) and the measured results, and this is also true for (47) albeit to a lesser degree of congruence for the higher $2B_L \cdot T$ factors. It should be noted, however, that the values of $2B_L \cdot T = 0.1$ and $2B_L \cdot T = 0.25$ are extremely high compared to those usually found in carrier synchronization PLLs; typically $2B_L \cdot T$ is in the order of magnitude of 0.01 at most ([25], [58], [22 Chap. 5, 6], [21 Chap. 5]), for which (as Fig. 31 illustrates) the values obtained through (30) and (44) (and even (47)) are virtually identical for all practical E_S / N_0 ratios. The rather large $2B_L \cdot T$ values of 0.1 and 0.25 in Fig. 32 and Fig. 33, respectively, were used with the intent of making the plots distinguishable, so that the reader could ascertain a qualitative appreciation of the influence of carrier phase jitter on the lock metric value. Consequently, it can be said that for most practical carrier synchronization PLLs the effects of carrier phase jitter on the lock detector value can be neglected.

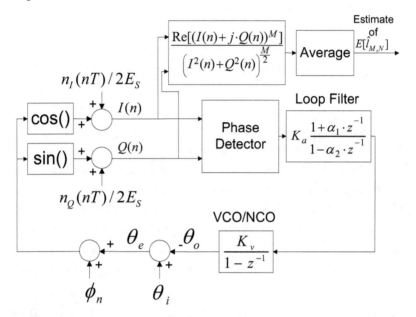

Fig. 30. Equivalent baseband model of the synchronization loop that was used for closed loop simulations.

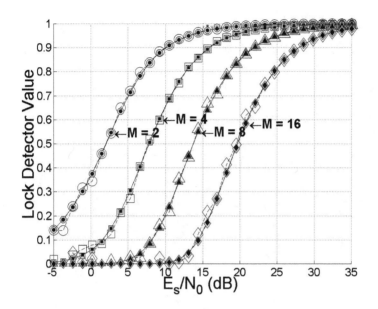

Fig. 31. Theoretical and simulated lock detector values, using equivalent baseband model of Fig. 30, for $2B_L \cdot T = 0.01$. Solid line is the theoretical, jitter free case (Eq. (30)). Blank polygons connected by dashed lines are the averages of measured lock detector values obtained in the simulation. Gray-filled polygons are approximations for Eq. (44), i.e. the results of multiplying Eq. (30) by an estimate (which we will denote $\hat{v}(\chi)$), of $v(\chi) \triangleq E\left[\cos\left(M\theta_e\right)\middle| E_S/N_0 = \chi\right]$, obtained by the time average $\hat{v}(\chi) = \overline{\cos(M\theta_e)}$, where θ_e was measured in the simulation. Similarly, black-filled polygons connected by a dashed line are approximations to Eq. (47), where $\hat{v}(\chi) = 1 - \frac{1}{2}M^2 \cdot \overline{\theta_e^2}$.

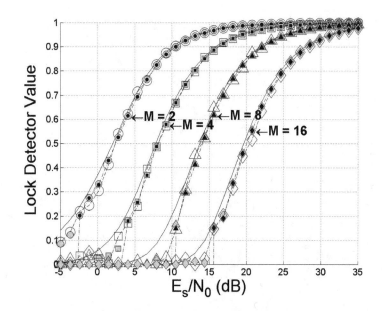

Fig. 32. Theoretical and simulated lock detector values, using equivalent baseband model of Fig. 30, for $2B_L \cdot T = 0.1$. Solid line is the theoretical, jitter free case (Eq. (30)). Blank polygons connected by dashed lines are the averages of measured lock detector values obtained in the simulation. Gray-filled polygons are approximations for Eq. (44), i.e. the results of multiplying Eq. (30) by an estimate (which we will denote $\hat{\upsilon}(\chi)$), of $\upsilon(\chi) \triangleq E\left[\cos\left(M\theta_e\right)\middle| E_S / N_0 = \chi\right]$, obtained by the time average $\hat{\upsilon}(\chi) = \overline{\cos(M\theta_e)}$, where θ_e was measured in the simulation. Similarly, black-filled polygons connected by a dashed line are approximations to Eq. (47), where $\hat{\upsilon}(\chi) = 1 - \frac{1}{2}M^2 \cdot \overline{\theta_e^2}$.

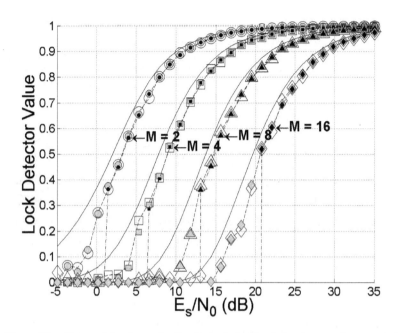

Fig. 33. Theoretical and simulated lock detector values, using equivalent baseband model of Fig. 30, for $2B_L \cdot T = 0.25$. Solid line is the theoretical, jitter free case (Eq. (30)). Blank polygons connected by dashed lines are the averages of measured lock detector values obtained in the simulation. Gray-filled polygons are approximations for Eq. (44), i.e. the results of multiplying Eq. (30) by an estimate (which we will denote $\hat{\upsilon}(\chi)$), of $\upsilon(\chi) \triangleq E\left[\cos\left(M\theta_e\right)\middle| E_S / N_0 = \chi\right]$, obtained by the time average $\hat{\upsilon}(\chi) = \overline{\cos(M\theta_e)}$, where θ_e was measured in the simulation. Similarly, black-filled polygons connected by a dashed line are approximations to Eq. (47), where $\hat{\upsilon}(\chi) = 1 - \frac{1}{2} M^2 \cdot \overline{\theta_e^2}$.

2.5   Lock Detector's Variance and Distribution

In order to allow for detection probabilities to be evaluated, the variance and distribution of the lock metrics must be ascertained. Recalling (17) and (18), since it has been established in Sec. 2.3.3 that $\forall M, \forall n, |x_{M,n}| \leq 1$ we consequently arrive at the bound:

$$\text{var}(x_{M,n}) = E[x_{M,n}^2] - E^2[x_{M,n}] \leq \left(\sup |x_{M,n}|\right)^2 = 1 \tag{48}$$

If I and Q sampling is done at the ideal instants (i.e. the symbol synchronization loop is ideally locked) then the symbol components of those samples are mutually independent, which is also true regarding the noise components (due to the matched filters). Thus the variables $x_{M,n}$ may be viewed as mutually independent, and hence:

$$\text{var}(\hat{l}_{M,N}) = \left(\frac{1}{2N}\right)^2 \sum_{n=-N+1}^{N} \text{var}(x_{M,n}) \leq \left(\frac{1}{2N}\right)^2 2N \cdot 1 = \frac{1}{2N} \tag{49}$$

Note that since no limiting assumptions were made, (49) is valid for any input signal at any E_S/N_0, including a noise-only signal, and for any carrier synchronization loop. In particular, (49) is valid for any carrier phase jitter conditions; the value of this observation will be apparent in Section 2.6. Due to the central limit theorem [13 Sec. 2.1.6] the distribution of $\hat{l}_{M,N}$ can be considered Gaussian when in lock since it is a sum of independent equally distributed variables $x_{M,n}$, and this is also true for the unlocked case provided there is no significant frequency error between the local and received carriers. Thus to a good (if conservative in terms of variance) approximation:

$$\hat{l}_{M,N} \mid \text{locked} \sim N\left(\mu_L(\chi), \frac{1}{2N}\right)$$

$$\hat{l}_{M,N} \mid \text{unlocked or noise only input} \sim N\left(0, \frac{1}{2N}\right) \tag{50}$$

where for jitter-free locking $\mu_L(\chi) \triangleq f_M(\chi)$ (given in (30)-(32)) and if modeling of the effects of the carrier synchronization PLL's phase jitter is deemed necessary, $\mu_L(\chi) \triangleq \tilde{f}_M(\chi)$ (given by (44)), as was discussed in Section 2.4.

2.6   Lock Probabilities and Circuit Parameter Determination

Eqs. (50) and (30)-(32), (44) can be used to set the threshold for achieving lock, from which it is evident that (with M assumed an unalterable system-level constant) the only quantity that needs to be decided upon by the designer in order to facilitate determination of the lock detector circuit's parameters is the minimum E_S / N_0 for which reliable lock is desired. To illustrate this, for a given threshold $\Gamma > 0$, at a given input E_S / N_0 ratio that we will denote χ, the lock detection probability P_D and false alarm probability P_{FA} are from (50) (assuming[5] $\Gamma < \mu_L(\chi)$, which is reasonable since otherwise we wouldn't be interested in detecting lock at an E_S / N_0 of χ):

$$P_D = P\left(\hat{l}_{M,N} > \Gamma \left| \frac{E_s}{N_0} = \chi \right. \right)$$

$$\geq \frac{1}{\sqrt{\pi / N}} \int_{\Gamma}^{\infty} e^{-(\tau - \mu_L(\chi))^2 \cdot N} d\tau = \frac{1}{2} erfc\left(\sqrt{N} \left(\Gamma - \mu_L(\chi) \right) \right)$$

(51)

$$P_{FA} = P\left(\hat{l}_{M,N} > \Gamma \left| noise\ only\ input \right. \right)$$

$$\leq \frac{1}{\sqrt{\pi / N}} \int_{\Gamma}^{\infty} e^{-\tau^2 \cdot N} d\tau = \frac{1}{2} erfc\left(\sqrt{N} \cdot \Gamma \right)$$

(52)

Solving (51) and (52) through a series of straightforward manipulations yields that suitable[6] values of N and Γ are:

---

[5] In fact we are implicitly making the assumption here that we are disinterested in detecting lock for *any* $E_S / N_0 = \lambda$ for which $\Gamma \geq \mu_L(\lambda)$; this is because for all $E_S / N_0 = \lambda$ for which $\Gamma \geq \mu_L(\lambda)$ we have $P_D \leq \frac{1}{2}$.

[6] Note that (51) and (52) are inequalities due to the fact that the variance of $\hat{l}_{M,N}$ is bounded by the limit $1/(2N)$ (see (49)), and not necessarily equal to it. Thus computation of (53) will produce somewhat conservative (i.e. larger than needed) values of N.

$$N = \left(\frac{erfc^{-1}\left(2P_{FA}\right) - erfc^{-1}\left(2P_D\right)}{\mu_L(\chi)} \right)^2 \tag{53}$$

and:

$$\Gamma = \frac{erfc^{-1}\left(2P_{FA}\right) \cdot \mu_L(\chi)}{erfc^{-1}\left(2P_{FA}\right) - erfc^{-1}\left(2P_D\right)} \tag{54}$$

Unsurprisingly, (51)-(54) are completely independent of K. As has been noted in Section 2.5, since $1/(2N)$ is an absolute upper bound on the variance of $\hat{l}_{M,N}$ which is valid for any phase jitter conditions, eqs. (51)-(54) are equally applicable if significant phase jitter is present, so long as $\mu_L(\chi) \triangleq \tilde{f}_M(\chi)$ (given in (44) or approximated by (47)).

Curves of (54) are shown in Fig. 34 for various values of M. Note that the value of χ chosen for each M corresponds to a reasonable minimal SNR for which reliable lock can be expected for that modulation [58 Sec. III]. Graphs of (53) are given in Fig. 35 for QPSK and $\chi=1$ dB, for jitter free conditions and for a loop SNR of $\rho = 16$ dB, where ρ is defined as ([25 Sec. 3.1], [51]) $\rho \triangleq 1/E[\theta_e^2]$ and (47) is used to compute the graphs pertaining to $\rho = 16$ dB. The use of $\chi=1$ dB and $\rho = 16$ dB allows direct comparison of Fig. 35 to [51 Fig. 6], which, if undertaken, shows that the number of symbols (which is $2 \cdot N$) needed for $\hat{l}_{4,N}$ is somewhat larger (but not excessively so) than that required for the 4th-order nonlinearity lock detector $LNDA_{4,N}$ that is discussed in [51]. However it must be remembered that (in contrast to this book) the analysis in [51] makes the assumption generally made in previous analyses of lock detectors, namely that the AGC operates perfectly. As already noted, if the AGC in [51] is not ideal (and none ever is) this will likely have adverse effects on P_D and P_{FA}, which are unaccounted for in [51 Fig. 6]. Particularly, if an abrupt fading of the input signal is experienced, P_D and P_{FA} will be severely affected until the AGC has settled; because the AGC generally has a time constant that is several orders of magnitudes larger than the symbol interval [49 Chap. 7], it will thus take many symbol intervals for the circuit in [51] to operate anew at the required P_D and P_{FA}. Such a phenomenon is entirely absent[7] for $\hat{l}_{M,N}$, and this must be held in context when comparing [51 Fig. 6] to Fig. 35. Furthermore, it is worthwhile noting that the author of this book has re-simulated [51 Fig. 6] using the data provided in [51] and the result is shown in Fig. 36, which shows some discrepancy with [51 Fig. 6]. If the results of Fig. 35 are compared to Fig. 36, we see that $\hat{l}_{4,N}$ has almost identical (only very slightly worse) performance (in terms of required symbols) as $LNDA_{4,N}$.

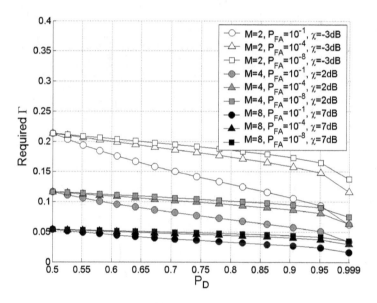

Fig. 34. Required threshold Γ, needed to achieve a necessary P_D and P_{FA}, for various values of M.

---

[7] So long as no signal-chain or sampler saturation occurs, and the samplers are not underdriven to the extent that quantization noise becomes significant. See Section 2.3.4.

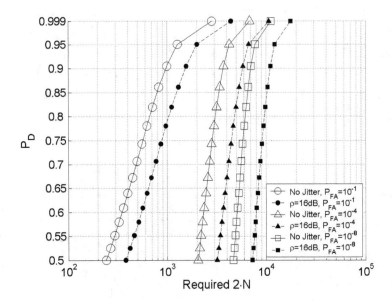

Fig. 35. Required No. of symbols, $2 \cdot N$, if using $\hat{l}_{4,N}$, needed to achieve a necessary P_D and P_{FA}, for $\chi = 1\,\mathrm{dB}$, for $M=4$ (QPSK), for jitter-free conditions and for a loop-SNR of $\rho = 16\,\mathrm{dB}$.

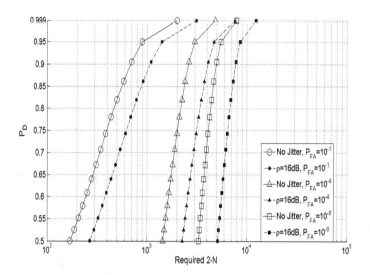

Fig. 36. Required No. of symbols, $2 \cdot N$**, if using** $LNDA_{4,N}$**, needed to achieve a necessary** P_D **and** P_{FA}**, for** $\chi = 1\,dB$**, for** $M=4$ **(QPSK), for jitter-free conditions and for a loop-SNR of** $\rho = 16\,dB$**.**

2.7 Operation in Fading Conditions

Up until this section we have assumed, rather conveniently, that the E_S / N_0 ratio is constant. In many cases, however, the signal experiences fading ([75], [76], [13 Chap. 14], [77]) which must be taken into account when modeling the lock detector's behaviour.

Before we engage in any quantitative discussions, we must first qualify the fading process under consideration. In the analysis of our lock detector, as we shall shortly see, we must differentiate between two cases according to the *channel coherence time* T_{COH}.

The channel coherence time is a measure of how fast the channel's characteristics are changing (see [13 Chap. 14]) . In general, we can assume that the channel E_S / N_0 is the same during time intervals which are significantly shorter than T_{COH} . Note that T_{COH} is the inverse of the channel's Doppler spread, that is $f_D = 1/T_{COH}$ (see [13 Sec. 14.1.1]). This is an important observation since in many papers the performance is given as a function of f_D, whereas here we have found it more convenient to work with T_{COH} , but since $f_D = 1/T_{COH}$ these measures are equally applicable.

In this book we assume that the fading is frequency-flat (i.e. $1/T \ll W_{COH}$ where W_{COH} is the coherence bandwidth of the channel [13 Chap. 14]) and slow (i.e. $T_{COH} \gg T$). The logic behind this choice is that we are dealing with single carrier coherent M-PSK system. If frequency-selective fading is present one would perhaps have chosen to implement a multicarrier communications system ([78], [79]), e.g. OFDM (Orthogonal Frequency Division Multiplexing). If fast fading had been present (i.e. $T_{COH} \ll T$), coherent demodulation without a pilot signal or pilot symbols would be very difficult, and hence one would have likely opted for a non-coherent or a differential modulation (see [13 p. 818], [78], [79]). Hence, the fact that the designer has chosen to implement a coherent, suppressed carrier, pilot-free M-PSK system almost assures, in and of itself, that the fading is frequency flat and slow. In this section we ignore the effects of phase jitter. This is permissible because of the analysis in Sec. 2.4 which showed that ignoring jitter-induced effects is a good assumption even at low SNRs.

The conditional probability distribution of the SNR due to fading will be denoted as $\mathbb{F}\left(\overline{\chi},\sigma_F{}^2\right)$ where $\overline{\chi}$ is the *average* SNR ratio (defined as $\overline{\chi} \triangleq E[E_S / N_0]$) and the associated pdf (probability density function) is denoted by:

$$p_F\left(\chi|\overline{\chi}\right) \triangleq p\left(E_S / N_0 = \chi \big| E[E_S / N_0] = \overline{\chi}\right) \tag{55}$$

We now present some common phase fading statistics, as taken from [77 Table 2]:

Table 2. Fading distributions for common channel types

Channel Type	Fading Distribution $p_F\left(\chi\middle|\overline{\chi}\right) \triangleq p\left(E_S/N_0 = \chi\middle|E[E_S/N_0] = \overline{\chi}\right)$
Rayleigh	$\dfrac{1}{\overline{\chi}}\exp\left(\dfrac{-\chi}{\overline{\chi}}\right)$
Rice (Nakagami-n)	$\dfrac{\left(1+n^2\right)e^{-n^2}}{\overline{\chi}}\exp\left(\dfrac{-\left(1+n^2\right)\chi}{\overline{\chi}}\right)I_0\left(2n\sqrt{\dfrac{\left(1+n^2\right)\chi}{\overline{\chi}}}\right)$
Hoyt (Nakagami-q)	$\dfrac{\left(1+q^2\right)e^{-n^2}}{2q\overline{\chi}}\exp\left(\dfrac{-\left(1+q^2\right)^2\chi}{4q^2\overline{\chi}}\right)I_0\left(\dfrac{\left(1-q^4\right)\chi}{4q\overline{\chi}}\right)$
Nakagami-m	$\dfrac{m^m\chi^{m-1}}{\overline{\chi}^m\Gamma(m)}\exp\left(\dfrac{-m\chi}{\overline{\chi}}\right)$

To evaluate the lock detector's behaviour during fading, we must differentiate between two cases[8]:

(a) $2N\cdot T \ll T_{COH}$: The lock detector calculation interval is much shorter than the coherence time. Thus, the E_S/N_0 can be considered constant during the lock detection computation process, and the analysis undertaken so far is applicable, regardless of the fading distribution.

(b) $2N\cdot T \gg T_{COH}$: The lock detector calculation interval is much longer than the coherence time. Thus, the effects of fading must be taken into account in computation of the lock metric.

---

[8] A third possibility, namely that $2N\cdot T \simeq T_{COH}$ (where "$\simeq$" means the same order of magnitude), is undesirable and can be avoided, as discussed in Sec. 2.7.1.

We now proceed to analyze each of these two cases.

2.7.1 Lock detector distribution for Case (a): $2N \cdot T << T_{COH}$

In case (a), the lock detector value will be based upon the *instantaneous* E_S / N_0, which we shall denote χ. In lieu of the fact that $2N \cdot T \ll T_{COH}$ this SNR can be considered constant during the lock detector computation process. Hence, the lock detector expected value will be as given in (30), i.e.:

$$f_M(\chi) \triangleq E\left[\hat{I}_{M,N}\Big|\frac{E_S}{N_0} = \chi\right] = \int_{-\pi}^{\pi} \cos(M\Delta\phi) p_R(\Delta\phi|\chi) d(\Delta\phi) \qquad (56)$$

where from (29):

$$p_R(\Delta\phi|\chi) \triangleq p(\Delta\phi_n = \Delta\phi | E_S / N_0 = \chi)$$

$$= \frac{\exp(-\chi)}{2\pi} \times \left[1 + \sqrt{2\chi}\cos(\Delta\phi)\exp(\chi \cdot \cos^2(\Delta\phi)) \cdot \int_{-\infty}^{\cos(\Delta\phi)\sqrt{2\chi}} e^{-y^2/2} dy\right] \qquad (57)$$

Similarly, it is trivial to note that the analysis given in Sections 2.3 to 2.6 is applicable without modification.

2.7.2 Lock detector distribution for Case (b): $2N \cdot T >> T_{COH}$

For case (b), since $2N \cdot T \gg T_{COH}$ then we can make two observations. First, since the lock detector estimation period is longer than the channel coherence time, then in the course of computing the lock metric we can expect to encounter changes in the SNR, and hence these must be taken into account when predicting the lock detector's performance. Secondly, since the estimation time is in fact *much* larger than the channel coherence time, we can assume that the E_S / N_0 values encountered during the lock detector computation time are distributed according to the channel fading statistics around the average SNR $\bar{\chi}$. Hence, it is expected (and, indeed, is verified shortly) that the lock detector value will be dependent upon two things: (i) the fading statistics, and (ii) the average SNR $\bar{\chi}$. Consequently, SNR estimation and threshold determination for this case will also be dependent upon $\bar{\chi}$ and the fading statistics.

Before we continue to quantitatively evaluate the performance of the lock detector for the case of $2N \cdot T \gg T_{COH}$, we comment that the case $2N \cdot T \sim T_{COH}$ (where " $\sim$ " means the same order of magnitude) is undesirable since in that case the SNR distribution during the lock detector computation period cannot be accurately predicted. This is because since $2N \cdot T \approx T_{COH}$ we are not statistically guaranteed that during the lock detector computation period of $2N \cdot T$ seconds the distribution of the SNR values encountered will be a sufficiently accurate approximation of $p_F(\chi | \overline{\chi})$, nor are we guaranteed that the SNR remains constant. Thus, for the case of $2N \cdot T \approx T_{COH}$, the lock-detector's value cannot be predicted and, *ipso facto*, the lock detection algorithm is rendered useless. Fortunately, $2N \cdot T \approx T_{COH}$ can always be avoided by choosing a large enough N, which ensures that $2N \cdot T \gg T_{COH}$ (case (b)). Ideally, though, we would prefer to always have $2N \cdot T \ll T_{COH}$ (case (a)), so that we could rapidly generate lock detector values and also maintain the ability to generate estimates of the instantaneous SNR (if we need the average SNR for some purpose, this can always be obtained by averaging the instantaneous SNR estimates). However, if the fading is such that T_{COH} is small, it might not be possible to find N which satisfies both the estimator accuracy requirements and yet which is small enough so that $2N \cdot T \ll T_{COH}$ is still obeyed. In that case, we must chose N such that $2N \cdot T \gg T_{COH}$ and content ourselves with a longer lock detector calculation period (which would be dependent upon (and produce an estimate of) the average SNR ratio $\overline{\chi}$).

Since $2N \cdot T \gg T_{COH}$, we have that the lock detector expectation is weighed by the SNR distribution given by the fading probability function, i.e.:

$$
\overline{f}_M(\overline{\chi}) \triangleq E\left[\hat{l}_{M,N} \middle| E\left[\frac{E_S}{N_0} \right] = \overline{\chi} \right]
$$

$$
= \int_0^\infty f_M(\chi) p_F(\chi | \overline{\chi}) d\chi \tag{58}
$$

$$
= \int_0^\infty \left(\int_{-\pi}^{\pi} \cos(M \Delta\phi) p_R \left(\Delta\phi \middle| \frac{E_S}{N_0} = \chi \right) d(\Delta\phi) \right) p_F(\chi | \overline{\chi}) d\chi
$$

Using (31) we may also write:

-69-

$$\overline{f}_M(\overline{\chi}) = \int_0^\infty \frac{\sqrt{\pi \cdot \chi}}{2} \cdot \exp\left(\frac{-\chi}{2}\right)\left[I_{\frac{M-1}{2}}\left(\frac{\chi}{2}\right) + I_{\frac{M+1}{2}}\left(\frac{\chi}{2}\right)\right]p_F\left(\chi|\overline{\chi}\right)d\chi \quad (59)$$

Regarding the lock detector's variance and distribution, we note, first, that the assumptions that led to the bound $\mathrm{var}\left(\hat{l}_{M,N}\right) \leq \frac{1}{2N}$ in Section 2.5 are still valid. Such is also the case regarding the conditions outlined in Section 2.5 which led to the conclusion that $\hat{l}_{M,N}$ is Gaussian. Hence, we conclude that (50) is still valid for the case of fading discussed here, provided that we use $\mu_L(\overline{\chi}) \triangleq \overline{f}_M(\overline{\chi})$ is as given by (58) instead of $\mu_L(\chi)$.

Secondly, as an immediate consequence of the previous paragraph we conclude that the analysis given in Section 2.6 is still valid, provided that we use $\mu_L(\overline{\chi}) \triangleq \overline{f}_M(\overline{\chi})$ as given by (58) instead of $\mu_L(\chi)$.

As a consequence of the above analysis, we conclude that in order to apply the equations in Sections 2.3-2.6 all we need to do is to compute $\overline{f}_M(\overline{\chi})$ via (58) and then use this in the appropriate equations. Obviously, there will be as many lock detector distributions as there are fading probability distributions, namely such analysis gives rise to an infinite number of lock detector expectation curves. Rather than attempt the impossible task of presenting results for all possible fading distributions, in the sequel we shall present some specific results for Nakagami-m fading, with the understanding that computations for other fading statistics can be done in an analogous manner.

2.7.3   Example: Operation in Nakagami-*m* fading for *2N·T >>T_COH*

In this section we shall evaluate, as an example, the lock detector's operation in the presence of Nakagami-m fading $2N \cdot T \gg T_{COH}$. As outlined in previous subsection, the distribution of the lock detector is still Gaussian, and the bound $\mathrm{var}\left(\hat{l}_{M,N}\right) \leq \frac{1}{2N}$ is still valid. Hence, all that remains to characterize is the lock detector's expected value, namely $\overline{f}_M(\overline{\chi}) \triangleq E\left[\hat{i}_{M,N} \middle| E\left[\frac{E_S}{N_0}\right] = \overline{\chi}\right]$. This has been done in Appendix B, where it was found that:

$$\overline{f}_M(\overline{\chi}) = \tilde{\Upsilon}_{0,0,m}(\overline{\chi}) + \frac{M}{2}\sum_{n=1}^{M/2}\frac{(-1)^n}{n!}\cdot\frac{(M/2+n-1)!}{(M/2-n)!}\cdot\tilde{\Upsilon}_{n,0,m}(\overline{\chi})$$
$$+\left[(-1)^{M/2+1}\sum_{n=1}^{M/2}\frac{1}{(n-1)!}\frac{(M/2+n-1)!}{(M/2-n)!}\tilde{\Upsilon}_{n,1,m}(\overline{\chi})\right]$$

(60)

where:

$$\tilde{\Upsilon}_{k,\ell,m}(\overline{\chi}) \triangleq \begin{cases} \Upsilon_{k,\ell,m}(\overline{\chi}) & \text{m-k} \notin \{0,-1,-2,-3,...\} \\ \Upsilon_{k,\ell,\tilde{m}}(\overline{\chi}) & \text{m-k} \in \{0,-1,-2,-3,...\} \end{cases}$$

(61)

where $k,\ell \geq 0$ and $m \geq 0.5$, and $\tilde{m} \triangleq m + m/1000$

and:

$$\Upsilon_{k,\ell,m}(\overline{\chi}) \triangleq \frac{\overline{\chi}^{-k}m^m \cdot \Gamma(m-k)}{(\ell\overline{\chi}+m)^{m-k}\Gamma(m)}$$

(62)

where $k,\ell \geq 0$ and $m \geq 0.5$

Curves of (60) are shown in Fig. 37. As seen in that figure, the lock detector curves have similar shapes in the presence of fading as compared to their values when operating in a fading-free environment. Note, furthermore, that results for $m = 1$ correspond to Rayleigh fading (see [13 Sec. 14.3]).

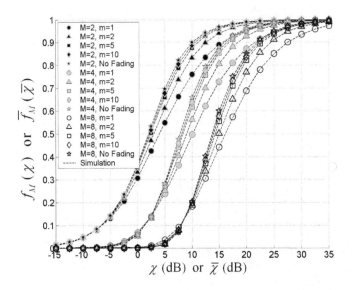

Fig. 37. Theoretical expectation of lock detector vs. simulations, for cases of (i) no fading (or, equivalently, fading but with $2N \cdot T \ll T_{COH}$), and (ii) for frequency-flat slow Nakagami-m fading with $2N \cdot T \gg T_{COH}$ for various values of m. For case (i) the graph shows $f_M(\chi)$ vs. χ, while for case (ii) the graph is of $\overline{f}_M(\overline{\chi})$ vs. $\overline{\chi}$. The simulations used $N=512$ and 100 lock metrics were averaged to compute each simulation data point. For the computation of $\overline{f}_M(\overline{\chi})$ the simulations used $T_{COH} = 10 \cdot T$, while for computation of $f_M(\chi)$ the SNR was assumed to be constant (i.e. no fading, or equivalently $T_{COH} = \infty$).

2.8 Conclusions

A family of carrier lock detectors for M-PSK receivers was presented, its theoretical properties analyzed, and simulations used to validate the results. It was found that the proposed lock metrics could be of substantial practical significance, as they lend themselves to simple hardware implementation and have easily bounded variance behaviour that, along with self-normalizing qualities, facilitate straightforward lock threshold determination and detection probability computation. It was found that the proposed lock detectors have a significant advantage over previously available lock detectors in terms of their ability to lend themselves to compact implementation in fixed-point hardware. Moreover, the AGC-independence afforded by the proposed detectors is also an advantage over previously available structures. In the final part of the chapter, the operation of the lock detector in the presence of fading was discussed, and specific results were given for Nakagami-m fading.

It was further found that the channel E_s / N_0 could be easily estimated from the lock metric value. This shall be investigated in depth in Chapter 4 Part A. We shall also see that the proposed lock detector has other advantages. In Chapter 3 we shall see that it can be used as a building block of an adaptive phase detector structure. In Chapter 4 Part B we shall use a modified lock detector structure in order to achieve SNR estimation for D-MPSK and for M-PSK in the absence of carrier synchronization. Thus, the lock detectors presented in this chapter are not only useful for lock detection, but have a range of other applications in SNR estimation and phase detection, issues that are explored further in the remainder of this book.

Chapter 3 Robust M-PSK Phase Detector Structures: Theory, Simulations, and System Identification Analysis

3.1 Introduction

Carrier phase error removal in M-PSK demodulators is generally achieved via one of two techniques. The first method uses a feedforward phase estimator ([22 Chap. 5, 6], [80], [81], [82], [83], [84], [85], [86], [87]) such as the Viterbi & Viterbi (V&V) detector [80] to estimate the phase error, and that estimate is then used to demodulate the received signal. The second method is the use of feedback ([22 Chap. 5, 6], [88], [89 Chap. 9], [90 Chap. 16], [91], [92], [93], [94], [95], [96]) systems, which remove the carrier phase error using a Phase Locked Loop (PLL) that ideally cancels the phase error between the local and received carriers. While the phase-error variance performances of appropriately chosen feedforward and feedback systems converge to the same bounds at high SNR (see [22 Chap. 6]), at moderate and low SNRs feedback systems exhibit nonlinear behavioural artefacts that cannot be modeled by an equivalent feedforward system.

A simplified illustration of feedforward demodulation of an M-PSK burst is shown in Fig. 38. Illustration of the topology of feedback receivers has already been discussed in Sec. 1.3, and diagrams of the place of the phase detector within such a system can be seen in Fig. 12 and Fig. 13.

This chapter was presented in part in Linn,Y.,"A Robust Phase Detection Structure for M-PSK: Theoretical Derivations, Simulation Results, and System Identification Analysis", 18th Canadian Conference on Electrical and Computer Engineering (CCECE'05), May 1-4, 2005, Saskatoon, SK, Canada, pp. 869-883. Some parts will also be published in "Robust M-PSK Phase Detectors for Carrier Synchronization PLLs in Coherent Receivers: Theory and Simulations", IEEE Transactions on Communications (in press).

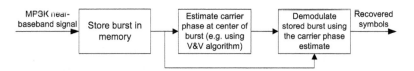

Fig. 38. Simplified flow diagram of feedforward demodulation of an M-PSK burst. Note that storage and carrier phase estimation can proceed concurrently.

It is instructive to engage in a point-by-point comparison of the qualities of feedback phase detectors with those of feedforward carrier phase estimators.

Phase detectors:

1. Are used in closed-loop as part of a PLL.

2. The receiver achieves coherent carrier synchronization (i.e. $\Delta\omega = 0$ and $\theta_e \approx 0$).

3. A phase error estimate is produced for each symbol.

4. The PLL that produces the phase estimate is an IIR (Infinite Impulse Response) system.

5. The phase estimate is causal.

6. There is an acquisition time for the PLL to acquire lock.

Feedforward carrier phase estimators:

1. Are used in an open-loop fashion and cannot be used in a PLL.

2. The receiver does *not* achieve coherent carrier synchronization (i.e. $\Delta\omega \neq 0$ and $\theta_e \neq 0$).

3. One phase error estimate is produced for each *block* of symbols.

4. The estimator produces a phase estimate using (in most cases) an FIR (Finite Impulse Response) system .

5. The phase estimate is non-causal.

6. There is no acquisition time.

Phase-detectors and feedforward carrier phase estimators have different advantages and disadvantages which make them suitable for different tasks. Phase detectors are used when the communications signal is continuous, which allows for acquisition and tracking of the input carrier by a PLL. In contrast, feedforward carrier phase estimators, such as the Viterbi & Viterbi estimator [80], are best suited for burst transmission which is too short to allow a carrier PLL to be employed because the acquisition latency cannot be tolerated. The tradeoff is that the amount of processing required for feedforward phase estimation is much larger and the rate of phase estimates is much reduced :one needs to store the entire burst in memory, and remove the phase error later after a single phase error estimate is made upon the burst; this is also a manifestation of the non-causality of the estimate. Increasing the rate of phase estimates garnered from a feedforward phase estimator requires even more significant processing, i.e. by using multiple overlapping estimation windows ([22 Sec. 6.5.4], [80]).

The tracking error variance of a feedback system will tend at high SNR to the same Cramér-Rao Bound of a feedforward system if we choose $N = 1/(2B_L T)$ where N is the number of symbols used to estimate the phase in the feedforward system, B_L is the noise bandwidth of the feedback system, and $1/T$ is the symbol rate (see [22 Chap. 6]). However, at moderate and low SNR, loop nonlinearities and/or decision errors will cause an increase in the phase error variance of the feedback system. This phenomenon will not be observed for the corresponding feedforward system. Put another way, at low SNRs the feedback phase detector operates more and more in a nonlinear capacity, the linearization assumptions of the PLL break down, the nonlinear behaviour of the PLL dominates, and this results in an additional degradation in the phase error variance. Even at moderate E_S / N_0, the phase detector will operate a large enough portion of the time in its nonlinear region, hence affecting the PLL dynamics and the phase-error variance. The feedforward and feedback systems cannot remain equivalent at low and moderate SNR, because the PLL's system dynamics change with decreasing SNR (behaving more and more as a nonlinear system), while the feedforward system's dynamics remains unchanged.

-76-

In this book, as already noted, we concentrate solely on coherent M-PSK receivers that utilize feedback in the form of a carrier synchronization PLL to remove the carrier phase error. Carrier synchronization PLLs in coherent M-PSK receivers are tasked with cancelling the carrier phase error, an estimate of which is provided by a carrier Phase Detector (PD). There are two general categories of PDs: Non Data Aided (NDA) and Decision Directed (DD). The Mth-order nonlinearity detector ([13 Chap. 6], [22 Chap. 5,6], [21 Chap. 5], [92]) and the multiphase NDA Costas loop ([26], [21 Chap. 5], [92]), and the multiphase Costa loop or M^{th}-order order nonlinearity followed by a limiter [92] are examples of NDA phase detectors. Examples of DD detectors can be found in [13 Chap. 6], [22 Chap. 5, 6], [58], [9], [97], [98], [93] and [91].

An inherent problem of DD detectors is that at low SNRs (and, as well, during acquisition) they suffer from considerable self-noise due to erroneous decisions, something which also has an effect on their S-Curves (see [22 Fig. 6-2], [58], [9], [97]). NDA detectors, while not susceptible to such a phenomenon, are nonetheless seldom used for higher order modulations. This is because at higher Ms NDA detectors experience high self-noise (due to the high-order nonlinearities which they include) and their implementations are significantly more complicated than their Decision Directed counterparts (see for example [26 p. 74], [21 Fig. 5.54]).

An additional problem which afflicts the DD and NDA detectors just cited is that their gain is strongly linked to the AGC circuit's operating point and performance. As we shall see in this chapter, the fact that the gain of the phase detector is not constant implies that (unless this change of gain is compensated for in some manner) the carrier PLL's characteristics will change accordingly. AGC-dependence is a particularly bothersome problem when fading signals are encountered, since in such cases the AGC often operates in a distinctly non-ideal manner, which means that AGC-dependence of the phase detector implies a similar lack of optimality in the carrier PLL.

This chapter's objective is the investigation of new families of M-PSK phase detectors. The first of those families will be a modification of the Mth-order nonlinearity detectors. The behaviour of the phase detectors is explored using stochastic theory, and it is found that the phase detectors' gain curves are independent of AGC performance and signal-levels. Furthermore, these self-normalizing properties allow the performance requirements of the AGC circuit to be relaxed considerably. Analysis of the squaring loss of the phase detectors is presented, and their closed-loop phase error variance is calculated as well as simulated and found to be better than that of Mth-order nonlinearity detectors and comparable to that of DD detectors. The phase detectors will also be shown to possess a simple hardware realization, which allows for their straightforward and efficient implementation within an FPGA or ASIC

The second family of new phase detectors will be a family of robust NDA adaptive phase detectors. These detectors will be shown to produce a constant-gain detector during tracking, which allows the carrier PLL to maintain optimality at virtually any SNR at which it can lock. The self-noise and phase-error variance performance of the proposed structure will be shown to be superior to that of other NDA and DD detectors. Moreover, unlike other NDA phase detectors, the proposed structure has a compact implementation for all M which is quite suitable for use within an FPGA or ASIC.

The organization of the chapter is as follows. In Section 3.2 we review the signal and receiver model around which the discussion applies, while in Section 3.3 we review the accepted metrics used in evaluation of phase detector performance. In Sections 3.4 we present a new family of self-normalizing NDA carrier phase detectors, where the characteristics and performance of this family of phase detectors are analyzed through theoretical derivations and simulations. In Sec. 3.5 we establish why a constant-gain phase detector is desirable, and in Section 3.6 we present a new family of adaptive M-PSK phase detectors which has such constant gain. The performance of these adaptive detectors is then investigated through theoretical derivations, simulations, and system-identification results. Finally, Section 3.7 is devoted to conclusions.

3.2   Signal and Receiver Models

The signal and receiver models, as well as the applicable notations, have been defined in Section 1.4.

3.3   Performance Analysis of PLLs – A Brief Overview

3.3.1   Performance metrics, nonlinear model, and linearized model

In this section we outline the PLL modeling techniques that we shall use to evaluate the proposed detectors. One of the most widely used PLL performance metrics [58 Sec I] is the *phase-error variance* $\text{var}(\theta_e)$, or equivalently, the *loop-SNR* defined[9] as [49 eq. (3.3-7)] $\rho \triangleq 1/\text{var}(\theta_e)$. This is because the phase-error variance has a crucial role in determining the cycle-slip rate of the PLL and the SER (Symbol Error Rate) degradation due to imperfect synchronization [21 p. 20-21, 210-211].

To understand this intuitively, we note that in order for the error rate to be small and to minimize the rate of cycle slips, we must have ([100], [99]) a small θ_e, i.e. we desire $\theta_e \approx 0$. Since θ_e is a zero-mean random process (when the PLL is locked), this means that for $\theta_e \approx 0$ to hold statistically we must have a small $\text{var}(\theta_e)$ or, equivalently, a high loop-SNR $\rho(= 1/\text{var}(\theta_e))$. Now, S-Curves of M-PSK phase detectors are periodic with period $2\pi/M$ (see for example Fig. 40, [22 Fig. 6-2]), so if the phase-error strays outside the range $\theta_e \in [-\pi/M, \pi/M]$ we will observe a loss of lock which is at least momentary (i.e. a *cycle slip* [22 Chap. 6], [49 Chap. 6])). Thus, the meaning of $\theta_e \approx 0$ for M-PSK receivers is more precisely expressed as $|\theta_e| \ll \pi/M$. Therefore, recalling again that when in lock θ_e is a zero-mean random process [26 Chap. 2], for good performance in an M-PSK receiver we must have:

$$\text{var}(\theta_e) \ll \pi^2/M^2 \tag{63}$$

---

[9] Sometimes the loop-SNR is defined as $\rho \triangleq 1/\text{var}(\theta_o)$ (e.g. [99]). However, if we want to evaluate the performance of the phase detector only, then we should assume that there is no input-carrier phase noise, in which case we have $\text{var}(\theta_e) = \text{var}(\theta_o)$ and the two definitions of loop-SNR coincide. This is the assumption made in this book (as well as in other texts [49], [22], [21], [9], [97], [58]) and we adopt the definition $\rho \triangleq 1/\text{var}(\theta_e)$.

It is clear from (63) that as M increases the requirements upon $\text{var}(\theta_e)$ will be more stringent, i.e. a higher loop-SNR will be required in order to achieve good performance. Hence the useful E_S/N_0 operating range of the PLL will start at a higher E_S/N_0 as M increases.

Determination of $\text{var}(\theta_e)$ via simulations is easily done ([101 Chap. 5], [102 Sec. 7.6]) using nonlinear models (shown in Fig. 47, Fig. 60), but in general the nonlinearity of the phase detector function presents great obstacles when *theoretical* analysis is attempted. To arrive at theoretical predictions, a standard approach adopted by synchronization texts (e.g. [22], [21], [21], [9], [97], [58]) has been to assume that the PLL is locked and then analyze the linearized PLL model. In the following subsections, we briefly review this approach.

3.3.2   The linearized PLL model

To develop the linearized model, we define the following quantities for any phase detector $P(n)$:

1) $B_L \triangleq \int_0^\infty \left| H_{PLL}(j2\pi f) \right|^2 df = \tfrac{1}{2}\omega_n(\zeta + 1/(4\zeta))$ is[10] the loop's *noise bandwidth* [25 p. 30-32].

2) The phase detector's *S-Curve* [21 p. 206] is $S_P(\theta_e) \triangleq E[P(n)|\theta_e]$, i.e. it is the average output of the phase detector given the phase error. Note that in general $S_P(\theta_e)$ will also have a dependence upon M, K, and χ, but for simplicity this is not indicated in the notation of the function $S_P(\theta_e)$.

3) The *linearized gain* of $P(n)$ (or simply the "*gain*" of $P(n)$) is:

$$g_P(M,K,\chi) \triangleq \left(\partial S_P(\theta_e)/\partial \theta_e \right)\Big|_{\theta_e=0} \tag{64}$$

---

[10] Note that this definition contains a factor of ½ w.r.t. the definition in [58]. This factor has been inserted in order to make the definition of B_L compatible with the (arguably) more widely used definition, as used throughout [21], [49], [22], and [25]. However, note that (69) incorporates a factor of 2 (w.r.t. the corresponding equation in [58]), so the aforementioned factor of ½ does not influence the phase-error variance results

Note in (64) that the gain generally depends on M, K, and χ. It is common practice to normalize the gain so that it is unity at $SNR - \infty$. Most synchronization texts also assume a constant $K = 1$, whereupon the normalized gain is:

$$\alpha_{\chi, P} \triangleq g_P(M, 1, \chi) / g_P(M, 1, \infty) \tag{65}$$

(note: since χ signifies SNR, we use the notations $\alpha_{SNR, P}$ and $\alpha_{\chi, P}$ interchangeably) $\alpha_{SNR, P}$ is called the *amplitude suppression factor*. This factor is *the multiplicative factor by which the expected linearized gain of the detector is reduced due to the presence of additive noise at its inputs, as compared to the phase detector's expected linearized gain for noiseless inputs.* However, as we showed in Sec. 1.5, despite its widespread use the assumption $K = 1$ is usually not realistic. Hence, in this chapter we assume that K is a function of the SNR, i.e. $K = \Upsilon_{AGC}(\chi)$ (see Sec. 1.5), and we define the *effective amplitude suppression factor*, denoted $\beta_{SNR, P}$, which (as we shall see later) is useful for incorporating AGC effects into the PLL model. The effective amplitude suppression factor for a given phase detector $P(n)$ is the *multiplicative factor by which the expected linearized gain of the phase detector is reduced due to the presence of additive noise **and variations of K** at its inputs, as compared to the phase detector's expected linearized gain for noiseless inputs **and K=1**.* Formally:

$$\beta_{\chi, P} \triangleq g_P(M, \Upsilon_{AGC}(\chi), \chi) / g_P(M, 1, \infty) \tag{66}$$

(note: since χ signifies SNR, we use the notations $\beta_{SNR, P}$ and $\beta_{\chi, P}$ interchangeably).

4) The normalized *equivalent loop noise* at $\theta_e \approx 0$ is defined as (using [22 eq. (6-73), p. 342] and normalizing)

$$N_{e, P}(n) \triangleq \lim_{\theta_e \to 0} (P(n) - S_P(\theta_e)) / g_P(M, 1, \infty). \tag{67}$$

5) The phase detector's *self noise* is defined as (see [58 eq. (6)])

$$\xi_P \triangleq 2 \cdot \chi \cdot \text{var}(N_{e, P}(n)). \tag{68}$$

-81-

Note that when AGC effects are ignored then (67) and (68) are computed with $K=1$ assumed. When AGC effects are modeled and $\beta_{SNR,P}$ is used as the gain in the linear model, then (67) and (68) **must** be computed using $K = \Upsilon_{AGC}(\chi)$.

6) An important tool in evaluating a phase detector is its *squaring loss* [13 eq. (6.2-59)]. In terms of previous definitions, if we assume $K=1$ the squaring loss is $\Omega_P \triangleq \xi_P / \alpha_{SNR,P}^2$, or $\Omega_P \triangleq \xi_P / \beta_{SNR,P}^2$ if modeling of the AGC's effects is done via $K = \Upsilon_{AGC}(\chi)$ (the squaring loss is identical in both cases[11], since it is an inherent property of the phase detector).

The linear model is shown in the lower left of Fig. 55. The phase-error variance is[12] [9 eq. (21)] :

$$\text{var}(\theta_e) = B_L \cdot T \cdot \Omega_P / \chi = \tfrac{1}{2}\omega_n(\zeta + 1/(4\zeta)) \cdot T \cdot \Omega_P / \chi \qquad (69)$$

3.4 A Family of Self-Normalizing Phase Detectors

In this section we shall endeavour to present and investigate a new family of self-normalizing carrier phase detectors for M-PSK receivers. We begin by defining the detector and then computing its S-Curve theoretically and via simulations. The amplitude suppression factors, self-noise, squaring loss, and phase-error variance performance are derived theoretically and then this is verified through nonlinear-model simulations. Finally, a compact hardware implementation for this family of detectors is presented, and the novelty of the proposed detectors is illustrated intuitively in a graphical fashion.

---

[11] It is important to note that when AGC effects are modeled via $\Omega_P \triangleq \xi_P / \beta_{SNR,P}^2$, then both denominator and numerator *must* be computed with $K = \Upsilon_{AGC}(\chi)$. When AGC effects are ignored, then $\Omega_P \triangleq \xi_P / \alpha_{SNR,P}^2$ where both denominator and numerator are computed with $K=1$. See Table 3.

[12] Note that the definition of B_L here contains a factor of ½ w.r.t. its definition in [9]. However, (69) compensates with a factor of 2 w.r.t. [9 eq. (21)].

3.4.1 Definition

Here we present a new family of M-PSK phase detectors. These can be thought of as a modification of the Mth-order nonlinearity detectors ([13], [22 Chap. 5, 6], [21 Chap. 5]) $c_M(n) \triangleq \text{Im}[(I(n) + j \cdot Q(n))^M]$. The modified detector is defined as follows:

$$d_M(n) \triangleq \frac{\text{Im}[(I(n) + jQ(n))^M]}{\left(I^2(n) + Q^2(n)\right)^{\frac{M}{2}}} = \frac{\sum\limits_{k=0}^{(M/2)-1} \binom{M}{2k+1} (-1)^k I^{M-2k-1}(n) Q^{2k+1}(n)}{\left(I^2(n) + Q^2(n)\right)^{\frac{M}{2}}}. \tag{70}$$

The denominator term in (70) performs adaptive normalization on the numerator, and - as we shall see - this normalization makes $d_M(n)$ behave quite differently from $c_M(n)$ (despite the notational resemblance). Another common phase detector that we shall use for comparisons is the decision directed detector [58] defined as $DD_M(n) \triangleq I(n) \cdot \hat{Q}(n) - Q(n) \cdot \hat{I}(n)$ (where $\hat{I}(n)$ and $\hat{Q}(n)$ are the I and Q decisions).

3.4.2 S-Curve of $d_M(n)$

The purpose of this section is to determine the S-Curve of the detector $d_M(n)$, we start by defining the phase of the received complex sample as $\varphi_n \triangleq \tan^{-1}\left(Q(n)/I(n)\right)$. Applying elementary rectangular-to-polar manipulations to (70) and using De Moivre's theorem[13] [54 eq. 6.9] yield:

$$d_M(n) = \frac{\text{Im}[(I(n) + jQ(n))^M]}{\left(I^2(n) + Q^2(n)\right)^{\frac{M}{2}}} = \frac{\left(I^2(n) + Q^2(n)\right)^{\frac{M}{2}} \sin\left(M\varphi_n\right)}{\left(I^2(n) + Q^2(n)\right)^{\frac{M}{2}}} = \sin\left(M\varphi_n\right). \tag{71}$$

When locked, $\Delta\omega = 0$ and $\theta_o \in \{\theta_i + 2\pi k/M - \theta_e | k = 0,1,...,M-1\}$, and hence from (21) φ_n reduces to:

---

[13] DeMoivre's theorem [54 eq. 6.9] states that for any real x and y, $(x + j \cdot y)^M = (x^2 + y^2)^{M/2} \exp(j \cdot M\theta)$ where $\theta \triangleq \tan^{-1}(y/x)$.

$$\varphi_n = \tan^{-1}\left(\frac{\sin(\theta_e + 2\pi(m_n - k)/M) + n_Q(nT)/(2E_S)}{\cos(\theta_e + 2\pi(m_n - k)/M) + n_I(nT)/(2E_S)}\right). \tag{72}$$

Coherent M-PSK systems overcome the carrier loop's ambiguity by differential precoding of the transmitted symbols [21 Sec. 5.7.6] or other Ambiguity Resolution (AR) circuits. We thus assume for convenience and without any loss of generality operation around the equilibrium point of $k = 0$. Furthermore, note that for any constellation point $\phi_x = 2\pi \cdot m_x / M$ (where m_x is an integer)

$$\sin(M(\varphi_n - \phi_x)) = \sin(M\varphi_n)\underbrace{\cos(2\pi \cdot m_x)}_{=1} - \cos(M\varphi_n)\underbrace{\sin(2\pi \cdot m_x)}_{=0} = \sin(M\varphi_n) . \tag{73}$$

Eqs. (71) and (73) thus shows that the output of $d_M(n)$ is not dependent upon the transmitted symbol but rather only on the phase difference between the received instantaneous symbol phase sample and the phase of *any* valid M-PSK constellation point. Thus we can assume for simplicity $\forall n$, $\phi_n = 0$, implying $\forall n$, $m_n = 0$. Under these simplifications we have when in lock, from (71) and (72):

$$d_M(n) = \sin\left(M \cdot \tan^{-1}\left(\frac{\sin(\theta_e) + n_Q(nT)/2E_S}{\cos(\theta_e) + n_I(nT)/2E_S}\right)\right). \tag{74}$$

Define the process $\Delta\phi_n \triangleq \varphi_n - \theta_e$. It can be shown (see [55 Sec. 4.5], as well as Sec. 2.4 of this book) that $\Delta\phi_n$ is distributed as the phase of a Rice random variable with a probability density function (pdf) for $|\Delta\phi| \le \pi$ of:

$$p_R(\Delta\phi \mid \chi) \triangleq p\left(\Delta\phi_n = \Delta\phi \mid E_S / N_0 = \chi\right)$$
$$= \frac{e^{-\chi}}{2\pi}\left[1 + \sqrt{2\chi}\cos(\Delta\phi)e^{\chi\cdot\cos^2(\Delta\phi)} \cdot \int_{-\infty}^{\cos(\Delta\phi)\sqrt{2\chi}} e^{-y^2/2}dy\right]. \tag{75}$$

To arrive at the S-Curve, recall that $\Delta\phi_n \triangleq \varphi_n - \theta_e$ so that $\varphi_n = \Delta\phi_n + \theta_e$ and thus from (71):

$$d_M(n) = \sin(M\varphi_n) = \sin(M(\Delta\phi_n + \theta_e))$$
$$= \cos(M\Delta\phi_n)\sin(M\theta_e) + \sin(M\Delta\phi_n)\cos(M\theta_e). \tag{76}$$

-84-

The S-Curve is then:

$$S_d(\theta_e) \triangleq E[d_M(n)|\theta_e]$$

$$= E[\cos(M\Delta\phi_n)]\sin(M\theta_e) + \underbrace{E[\sin(M\Delta\phi_n)]}_{=0}\cos(M\theta_e) \qquad (77)$$

$$= f_M(\chi)\sin(M\theta_e).$$

where $f_M(\chi) \triangleq \int_{-\pi}^{\pi} \cos(M\Delta\phi) \cdot p_R(\Delta\phi|\chi) \cdot d\Delta\phi$ was defined in (30) and investigated in Chapter 2. From (31) and (35) it was shown that:

$$f_M(\chi) = \frac{\sqrt{\pi \cdot \chi}}{2} \cdot \exp\left(\frac{-\chi}{2}\right)\left[I_{(M-1)/2}\left(\frac{\chi}{2}\right) + I_{(M+1)/2}\left(\frac{\chi}{2}\right)\right]$$

$$\overset{\text{high SNR}}{\approx} \exp\left(-M^2/(4\chi)\right) \qquad (78)$$

Substituting (78) into (77) yields the exact expression:

$$S_d(\theta_e) = \left(\frac{\sqrt{\pi \cdot \chi}}{2} \cdot \exp\left(\frac{-\chi}{2}\right)\left[I_{(M-1)/2}\left(\frac{\chi}{2}\right) + I_{(M+1)/2}\left(\frac{\chi}{2}\right)\right]\right)\sin(M\theta_e) \quad (79)$$

And the high-SNR approximation:

$$S_d(\theta_e) \overset{\text{high SNR}}{\approx} \exp\left(-M^2/(4\chi)\right)\sin(M\theta_e) \qquad (80)$$

We see from (79) that the S-Curve is **sinusoidal** with M stable equilibrium points, which is exactly what we would expect from an M-PSK phase detector. S-Curve plots are shown in Fig. 40, where we see that the approximation $\exp(-M^2/(4\chi))\sin(M\theta_e)$ is quite accurate (even at low SNR), hence making it a useful tool for the designer for manually estimating the S-Curve behaviour. The gain of d_M is from (64) and (79)-(80):

$$g_d(M,K,\chi) \triangleq \left(\partial S_d(\theta_e)/\partial\theta_e\right)\big|_{\theta_e=0} \cdot$$

$$= f_M(\chi) \cdot \left[\partial(\sin(M\theta_e))/\partial\theta_e\right]_{\theta_e=0} = M \cdot f_M(\chi)$$

$$= M \cdot \left(\frac{\sqrt{\pi \cdot \chi}}{2} \cdot \exp\left(\frac{-\chi}{2}\right)\left[I_{(M-1)/2}\left(\frac{\chi}{2}\right) + I_{(M+1)/2}\left(\frac{\chi}{2}\right)\right]\right) \qquad (81)$$

$$\overset{\text{high SNR}}{\approx} M \cdot \exp(-M^2/(4\chi))$$

3.4.3 Amplitude suppression factors of $d_M(n)$

The amplitude suppression factors are found from (65), (66), and (79)-(80):

$$\beta_{\chi,d} = \alpha_{\chi,d} = f_M(\chi) = \frac{\sqrt{\pi \cdot \chi}}{2} \cdot \exp\left(\frac{-\chi}{2}\right)\left[I_{(M-1)/2}\left(\frac{\chi}{2}\right) + I_{(M+1)/2}\left(\frac{\chi}{2}\right)\right]$$

$$\overset{\text{high SNR}}{\approx} \exp\left(-M^2/(4\chi)\right) \tag{82}$$

Exact closed-form expressions exist for $\alpha_{SNR,d}$ and $\beta_{SNR,d}$. These are easily obtained from (82) and (32):

$$\beta_{\chi,d} = \alpha_{\chi,d} = 1 + \frac{M}{2}\sum_{n=1}^{M/2}\frac{(-1)^n}{n!}\cdot\frac{(M/2+n-1)!}{(M/2-n)!\,\chi^n}$$

$$+ \exp\left(-\chi\right)\left[(-1)^{M/2+1}\sum_{n=1}^{M/2}\frac{1}{(n-1)!}\frac{(M/2+n-1)!}{(M/2-n)!\,\chi^n}\right] \tag{83}$$

We note that (83) is a <u>finite</u> sum composed of polynomials and exponentials, and hence is easily computed using numerical computation packages such as Matlab.

Graphs of (82) are given in Fig. 39. As can be seen in that figure, the approximation of $\exp\left(-M^2/(4\chi)\right)$ is quite accurate even at low SNR, particularly for $M>2$. Therefore, this is a useful approximation that can be used by the engineer for quick manual computations when designing the carrier PLL.

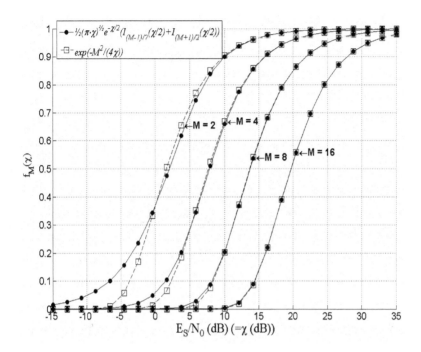

Fig. 39. Comparison of exact expression for $f_M(\chi)$ vs. the approximate expression. Note that since $\beta_{\chi,d} = \alpha_{\chi,d} = g_d / M = f_M(\chi)$ (see (81), (82)) then this figure is also useful for predicting $\beta_{\chi,d}$, $\alpha_{\chi,d}$, and g_d.

3.4.4   Loop noise, self noise, and squaring loss of $d_M(n)$

The loop noise is easily found from (76) and (67) to be :

$$N_{e,d}(n) = \frac{1}{M} \cdot \sin(M\Delta\phi_n) \tag{84}$$

Observe that $N_{e,d}$ is *not* Gaussian, though it may be approximated by Gaussian noise at high SNR, since (using (33)) we have $N_{e,d}(n) \xrightarrow{\chi \to \infty} \Delta\phi_n \overset{\text{high } \chi}{\sim} N(0, 1/(2\chi))$. The self noise of d_M is from (68), (84), (31) and (35):

$$
\xi_d = 2\chi \cdot \text{var}(N_{e,d}(n)) = 2\chi \cdot \int_{-\pi}^{\pi} \left(\frac{1}{M} \sin(M\tau) \right)^2 p_R(\tau \mid \chi) \cdot d\tau
$$

$$
= \frac{2\chi}{M^2} \int_{-\pi}^{\pi} \left(\frac{1}{2} - \frac{1}{2} \cos(2M\tau) \right) p_R(\tau \mid \chi) \cdot d\tau = \frac{\chi}{M^2} \left(1 - f_{2M}(\chi) \right)
$$

$$
= \frac{\chi}{M^2} \left(1 - \frac{\sqrt{\pi \cdot \chi}}{2} \cdot \exp\left(\frac{-\chi}{2} \right) \left[I_{(2M-1)/2}\left(\frac{\chi}{2} \right) + I_{(2M+1)/2}\left(\frac{\chi}{2} \right) \right] \right)
$$

$$
= \frac{\chi}{M^2} \left(\begin{array}{l} M \cdot \left[\displaystyle\sum_{n=1}^{M} \frac{(-1)^{n+1}}{n!} \cdot \frac{(M+n-1)!}{(M-n)! \chi^n} \right] \\[4mm] + \exp(-\chi) \left[(-1)^M \displaystyle\sum_{n=1}^{M} \frac{1}{(n-1)!} \frac{(M+n-1)!}{(M-n)! \chi^n} \right] \end{array} \right)
$$

(85)

and the squaring loss $\Omega_d = \xi_d / \alpha_{SNR,d}^2 \left(= \xi_d / \beta_{SNR,d}^2 \right)$ is readily found using (85) and (82).

3.4.5   Examples of S-Curves of $d_M(n)$

To give some graphical insight into the S-Curve of $d_M(n)$, we start by showing the complete S-Curve of $d_2(n)$, as shown in Fig. 40. In that figure, we have used (79) in order to plot the exact predicted values $S_d(\theta_e) = \left(\frac{\sqrt{\pi \cdot \chi}}{2} \cdot \exp\left(\frac{-\chi}{2} \right) \left[I_{(M-1)/2}\left(\frac{\chi}{2} \right) + I_{(M+1)/2}\left(\frac{\chi}{2} \right) \right] \right) \sin(M\theta_e)$, the S-Curve predicted by the approximation of (80) which is $S_d(\theta_e) \approx \exp\left(-M^2 / (4\chi) \right) \sin(M\theta_e)$, and simulation results (i.e. computation of $E[d_M(n)|\theta_e]$ through stochastic simulations). As can be seen from that figure, the simulation results agree perfectly with the exact

-88-

predicted results, and, furthermore, the approximation (80) is found to be an excellent approximation, even at low SNRs.

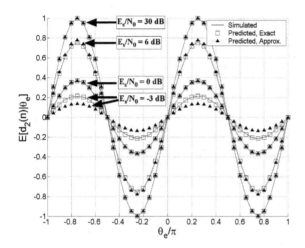

Fig. 40. S-Curve of $d_2(n)$ **for various SNR ratios ('Predicted, Exact' is** $S_d(\theta_e) = f_2(\chi)\sin(2\theta_e)$, **and 'Predicted, Approx.' is** $S_d(\theta_e) \approx \exp(-1/\chi)\sin(2\theta_e)$. **See (77), (79), (80)).**

To further give quantitative results, we first note that from (79) the S-Curves are periodic with period $\frac{2\pi}{M}$ over the entire interval $-\pi \le \theta_e \le \pi$. Hence, it suffices to plot the S-Curve over the interval $-\frac{\pi}{M} \le \theta_e \le \frac{\pi}{M}$. In Fig. 41 to Fig. 44 we see S-Curve examples for over the intervals $-\frac{\pi}{M} \le \theta_e \le \frac{\pi}{M}$ for M=2,4,8 and 16, for various SNR ratios. Plotted are the exact predicted S-Curves (eq. (79)), the approximated predicted S-Curves (eq. (80)), as well as simulation results (i.e., evaluation of $E[d_M(n)|\theta_e]$ through stochastic simulations of eq. (70)). We see from those figures that for M>2 the approximate

-89-

expression of (80) is almost a perfect match with the exact expression of (79). Hence, (80) is a useful tool which enables the designer to predict the S-Curve with accuracy. Note that the SNR values which are used to plot these figures are those which are appropriate for the given modulation, i.e. from low SNRs (= close to the lock threshold of the PLL for the respective modulation index M) through moderate SNRs and then to a high SNR value. For more information about lock thresholds the reader is referred to [58 Sec. III].

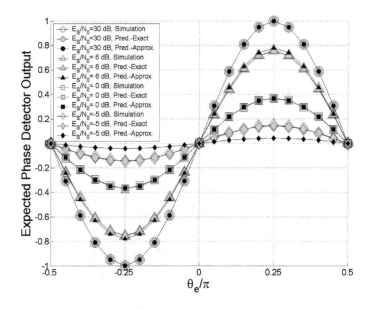

Fig. 41. S-Curve of $d_2(n)$ (for BPSK) for various SNRs in the interval $-\frac{\pi}{2} \leq \theta_e \leq \frac{\pi}{2}$. The phase detector curve is periodic with period $\frac{2\pi}{M} = \pi$ over the entire interval $-\pi \leq \theta_e \leq \pi$.

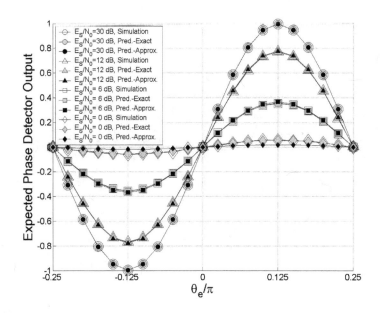

Fig. 42. S-Curve of $d_4(n)$ (for QPSK) for various SNRs in the
interval $-\frac{\pi}{4}\leq\theta_e\leq\frac{\pi}{4}$. The phase detector curve is periodic with
period $\frac{2\pi}{M}=\frac{\pi}{2}$ over the entire interval $-\pi\leq\theta_e\leq\pi$.

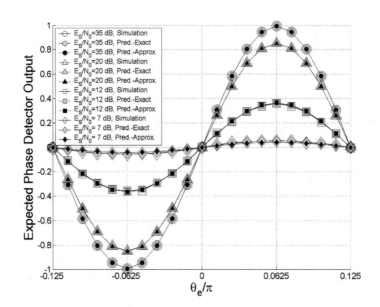

Fig. 43. S-Curve of $d_8(n)$ (for 8-PSK) for various SNRs in the interval $-\frac{\pi}{8} \le \theta_e \le \frac{\pi}{8}$. The phase detector curve is periodic with period $\frac{2\pi}{M} = \frac{\pi}{4}$ over the entire interval $-\pi \le \theta_e \le \pi$.

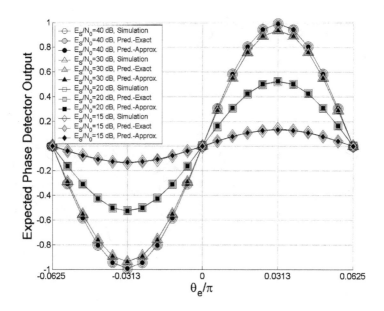

Fig. 44. S-Curve of $d_{16}(n)$ (for 16-PSK) for various SNRs in the interval $-\dfrac{\pi}{16} \le \theta_e \le \dfrac{\pi}{16}$. The phase detector curve is periodic with period $\dfrac{2\pi}{M} = \dfrac{\pi}{8}$ over the entire interval $-\pi \le \theta_e \le \pi$.

3.4.6   Squaring loss comparison vs. other phase detectors

Quantitative judgments regarding the squaring loss of d_M can be reached by comparing that loss to that which is incurred when c_M or DD_M is used. Such a comparison is shown in Fig. 45. Linear-model predictions of closed-loop phase-error variance of d_M can be made via (69) and are shown in Fig. 46, with $\zeta = 0.95$ and $\omega_n = 8.24 \cdot 10^{-3}/T$ fixed at all SNRs so that $2B_L T = 0.01$ at all SNR. The plots show that d_M

provides excellent performance, particularly compared to c_M. Hence, d_M is a very viable phase detector. Also plotted in Fig. 46 is the Cramér-Rao bound ([22], [21]), defined as:

$$CRB = B_L \cdot T / \chi = \frac{1}{2} \omega_n (\zeta + 1/(4\zeta)) \cdot T / \chi \qquad (86)$$

The data required to plot the results for c_M and DD_M was taken from Table 3.

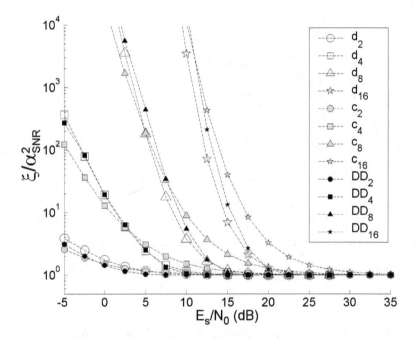

Fig. 45. Squaring loss as a function of E_s / N_0.

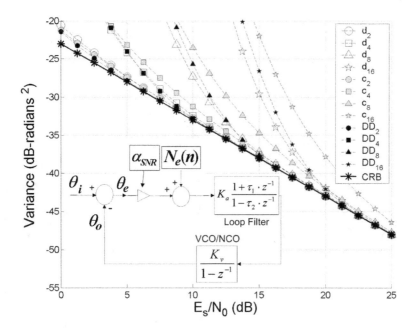

Fig. 46. Calculated phase-error variance $\text{var}(\theta_e)$, using linearized baseband model. The loop's noise bandwidth is held fixed at $2B_L \cdot T = 0.01$. AGC effects are ignored (i.e. $K=1$ identically). $\zeta = 0.95$.

3.4.7 Nonlinear model simulation investigation of $d_M(n)$.

To verify the linear-model predictions of Fig. 46, nonlinear simulations were conducted, with the simulation model and the results presented in Fig. 47. As we can see by comparing Fig. 46 to Fig. 47, the theoretical predictions of the linear model are in excellent agreement with the nonlinear simulation results anywhere above the PLL's lock threshold.

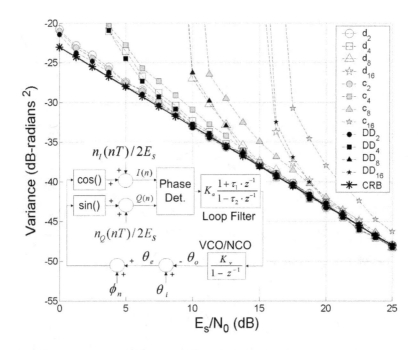

Fig. 47. Simulated phase-error variance $\text{var}(\theta_e)$, using equivalent baseband nonlinear model. Loop bandwidth is held fixed at $2B_L \cdot T = 0.01$. AGC effects are ignored (i.e. $K=1$ identically). $\zeta = 0.95$. The SNRs below which $\text{var}(\theta_e)$ increases dramatically for M=8 and M=16 are the *PLL lock thresholds* for those modulations.

3.4.8   Hardware realization

Since from (71) $d_M(n) = \sin(M\varphi_n)$, we see from inspection that $d_M(n)$'s value is independent of K, and hence independent of the AGC. Moreover, $d_M(n)$ has an efficient fixed-point hardware implementation in the form of a lookup table; this is due to exactly the same reasons that enabled such an implementation for $x_{M,n}$ in Section 2.3.3, namely the small dynamic range that is needed to express $d_M(n)$ (since $|d_M(n)| = |\sin(M\varphi_n)| \le 1$). The proposed implementation is shown in Fig. 48

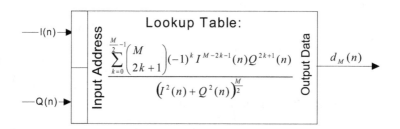

Fig. 48. Efficient hardware generation of $d_M(n)$

To see why the existence of such an implementation is significant, we make note of the fact that other phase detectors suffer from a large dynamic range that often renders a similar implementation unfeasible. To highlight this point, consider the Mth-order nonlinearity $c_M(n)$. It is easily seen that $c_M(n) \propto K^M$, which means that a phase detector lookup table and the ensuing datapath (in particular, the loop filter) must all be able to handle the dynamic range of K^M. This is often prohibitive to implement in fixed-point hardware. Moreover, the dependence on K^M implies a nonlinear dependence upon the AGC, a dependence that $d_M(n)$ does not exhibit. A similar conclusion can be reached with regards to Decision Directed detector ([58], [9], [97]) $DD_M(n) \triangleq I(n)\cdot\hat{Q}(n) - Q(n)\cdot\hat{I}(n)$ (where $\hat{I}(n)$ and $\hat{Q}(n)$ are decisions on the I and Q channels). Simple substitution shows that $DD_M(n) \propto K$, so use of $DD_M(n)$ means that a dependence upon the dynamic range of K and the AGC would still exist.

In contrast, for $d_M(n)$, the output of the lookup table is always in the interval $[-1,1]$, regardless of K or M. Thus, with $d_M(n)$, a fixed-point lookup table with just a 10-bit or even just an 8-bit output, i.e. quantization of the phase error estimate to 10 bits (quantization to 1024 levels of the interval $[-1,1]$) or 8 bits (quantization to 256 levels of the interval $[-1,1]$) will be more than enough for $d_M(n)$, for <u>any</u> K and <u>any</u> M.

Indeed, we see that d_M has many merits. In fact, it is an excellent M-PSK phase detector that can be used instead of c_M or DD_M. However, even better performance can be achieved by using d_M within an adaptive phase detection structure that is introduced in Section 3.6.

3.4.9   Lookup Table Implementation Issues

Though the lookup table's values of $d_M(n)$ are well behaved, a valid and necessary question is: what happens when both $I(n)$ and $Q(n)$ are 0? In that case the denominator of (70) vanishes, and the output of the corresponding lookup table is undefined. While in the unquantized theoretical analysis the event $I(n) = Q(n) = 0$ has an infinitesimal probability and is thus inconsequential, for the practical quantized case this eventuality has a finite probability and must be addressed.

Fortunately, there is an exceedingly simple way to solve this problem, which, quite fortuitously, also turns out to be the mathematically correct approach.

Let's consider for example BPSK. Then we have from (70) :

$$d_2(n) \triangleq \frac{2I(n)Q(n)}{I^2(n)+Q^2(n)} \tag{87}$$

Let's assume for the example that (87) is implemented in a lookup table with input quantization of 4 bits for each input, and the output of 8 bits, hence resulting in a $(2^4 \cdot 2^4) \cdot 8 = 2048$ bit lookup table. What happens when the I and Q inputs are both 0? Eq. (87) is useless in this case. However, the key here is to recognize that the quantization of the inputs to 4 bits means that we lack information regarding the bits that were not expressed in the quantized result. Assume that we treat all the binary numbers as fractions, i.e. that the binary point immediately follows the sign bit (this is a completely arbitrary convention). Then in 4-bit quantized binary form a zero value for the I and Q channels is $I = 0.000_2$ and $Q = 0.000_2$. Taking for example the I value, it is important to note that the value $I = 0.000_2$ will result from quantization of all true values of I between $I = (0.00000.....)_2 = 0$ and $I = (0.000111....)_2 = 0.125$. Thus, the correct way to interpret $I = 0.000_2$ is by the average of all possible values that it could represent, i.e. $(0.125 + 0)/2 = 0.0625 = 0.0001_2$. This reasoning applies not only to the 0-value case, but to all values. In other words, the mathematically correct way to interpret a quantized value is to think of the true value as one which contains an additional "invisible" LSB, whose value is 1. When the values of the lookup tables are computed with this value have the added advantage that the lookup table value for all-zero inputs is well-defined. For example, for the case of (87) with 4-bit inputs and an 8 bit output, we have for the 0-input case:

$$
\begin{aligned}
d_2(n)\big|_{\substack{I=0.000_2 \\ Q=0.000_2}} &\triangleq quant_8\left(\frac{2 \cdot 0.0625 \cdot 0.0625}{0.0625^2 + 0.0625^2}\right) \\
&= quant_8(1) = 0.1111111_2 (= 0.9921875)
\end{aligned}
$$

(88)

Contour maps of the corresponding lookup tables with and without quantization are shown in Fig. 50.

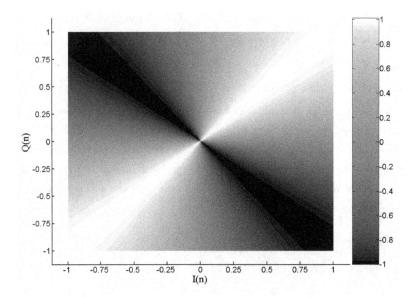

Fig. 49. Intensity graph for the lookup table computing $d_2(n) \triangleq 2I(n)Q(n)/(I^2(n)+Q^2(n))$, **no quantization**.

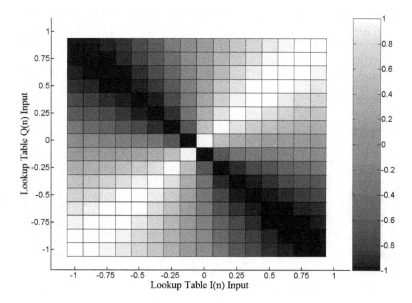

Fig. 50. Intensity graph for the lookup table computing $d_2(n) \triangleq 2I(n)Q(n)/(I^2(n)+Q^2(n))$, with quantization of inputs to 4 bits (each) and the output to 8 bits (note that the computation of the lookup table values is done as outlined in Sec. 3.4.9). The fact that the quantized I and Q ranges are asymmetric around 0 (they are [-1,0.875]) is due to the 2's complement representation.

3.4.10   Intuitive Understanding of $d_M(n)$

We shall now take an interlude from the rigorous mathematical procedure undertaken thus far, in order to attempt the instilment of an intuitive appreciation of the phase detectors' performance and novelty. This is facilitated by comparing $d_M(n)$ to the standard M^{th} order nonlinearity $c_M(n) = \text{Im}[(I(n) + j \cdot Q(n))^M]$.

As the chosen example we look at M=4 (QPSK). In Fig. 51 and Fig. 52 we see a contour map of $d_4(n) \equiv \dfrac{4I^3(n)Q(n) - 4I(n)Q^3(n)}{\left(I^2(n) + Q^2(n)\right)^2}$, upon which demodulated QPSK signals are superimposed, with $K = 0.3$ and $K = 0.8$, respectively. As can be clearly seen in the contour graph, the radial symmetry of the contours means that the phase detector output does not depend on the value of K; rather, it depends solely on the phase departure of the current received symbol from the ideal phase of a constellation point. This, however, is clearly not the case for the QPSK phase detector $c_4(n) = \text{Im}[(I(n) + j \cdot Q(n))^4]$, whose contour graph is featured in Fig. 53 and Fig. 54, upon which the same demodulated constellations are superimposed. As is evident from those figures, for the 4th order nonlinearity detector, there is a strong dependence of the phase detector value upon the value of K, and hence upon the AGC's performance and dynamic range.

The AGC-independence of $d_M(n)$ is not complete, in the sense that the AGC must still ensure that the samplers and preceding signal chains are not overdriven, and, conversely, that the samplers are not underdriven to the extent that quantization noise becomes significant. However, once the AGC meets these two basic requirements, the values of K (and the fluctuations thereof) are irrelevant.

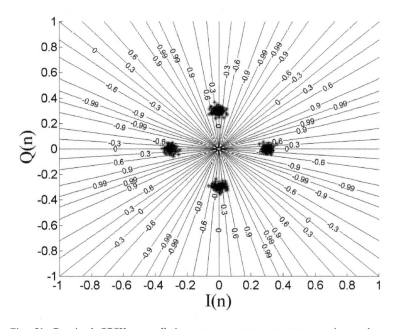

Fig. 51. Received QPSK constellation, $E_S / N_0 = 20dB$, $K = 0.3$, superimposed on a contour map of $d_4(n) = \dfrac{4I^3(n)Q(n) - 4I(n)Q^3(n)}{\left(I^2(n) + Q^2(n)\right)^2}$

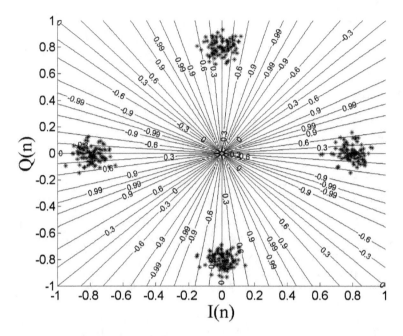

Fig. 52. Received QPSK constellation, $E_S/N_0 = 20dB$, $K = 0.8$, superimposed on a contour map of $d_4(n) = \dfrac{4I^3(n)Q(n) - 4I(n)Q^3(n)}{\left(I^2(n) + Q^2(n)\right)^2}$.

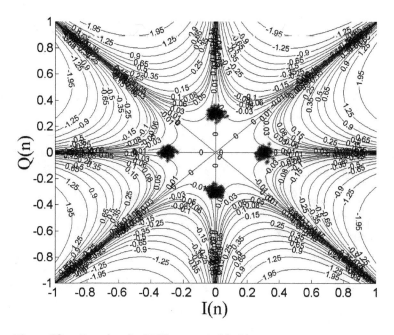

Fig. 53. Received QPSK constellation, $E_S/N_0 = 20dB$, $K = 0.3$, superimposed on a contour map of $c_4(n) = \mathrm{Im}[(I(n) + j \cdot Q(n))^4] = 4I^3(n)Q(n) - 4I(n)Q^3(n)$.

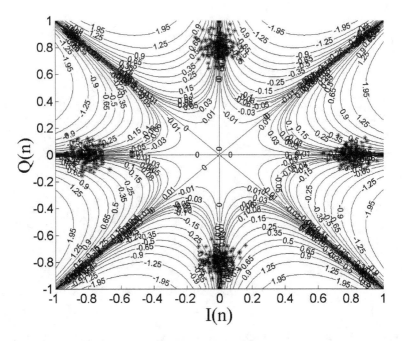

Fig. 54. Received QPSK constellation, $E_S / N_0 = 20dB$, $K = 0.8$, **superimposed on a contour map of** $c_4(n) = \text{Im}[(I(n) + j \cdot Q(n))^4] = 4I^3(n)Q(n) - 4I(n)Q^3(n)$.

Yet another way to interpret the functioning of $d_M(n)$ is by noticing that its amplitude

suppression factors $\beta_{\chi,d} = \alpha_{\chi,d} = \frac{\sqrt{\pi\chi}}{2} \cdot e^{-\chi/2} \left[I_{\frac{M-1}{2}}\left(\frac{\chi}{2}\right) + I_{\frac{M+1}{2}}\left(\frac{\chi}{2}\right) \right]$ are the same as those for a

limiter which follows (or precedes) an Mth-order nonlinearity (or, equivalently, a limiter

which follows (or precedes) a multiphase Costas phase detector), as is the case analyzed

in [92]. Since the amplitude suppression factors are the same, we can thus think of the

self-normalizing action of $d_M(n) \triangleq \dfrac{\text{Im}[(I(n) + jQ(n))^M]}{(I^2(n) + Q^2(n))^{M/2}}$ as that of the digital equivalent of a *limiter* (although, of course, the implementation of $d_M(n)$ is much more compact and suited for digital communications than the system considered in [92]).

3.5   The Case for a Constant-Gain Phase Detector

In Fig. 46 and Fig. 47 we have $2B_L T = \omega_n (\zeta + 1/(4\zeta)) \cdot T = 0.01$ at all SNRs. This is achieved by using a different loop-filter gain at each SNR so that $\zeta = 0.95$ and $\omega_n = 8.24 \cdot 10^{-3}/T$ (the loop-filter gain is the coefficient K_a in the loop-filter function. See the linear and nonlinear models in the lower left of Fig. 46, Fig. 47, and Fig. 60). Maintaining ζ and ω_n constant at all SNRs is the practice adopted in most synchronization texts since it facilitates a meaningful comparison of $\text{var}(\theta_e)$ achievable by the compared phase detectors when they are employed in PLLs that have identical parameters. However, it is important to note that constant ζ and ω_n <u>cannot</u> be maintained by a PLL with a *fixed* (=non-adaptive) loop-filter gain and which uses either DD_M, c_M, or d_M.

Consider, for example, in Figs. 46 and 47 the PLL for QPSK ($M=4$) which employs DD_M. Let us use the notation $K_{a,DD}(\rho)$ to refer to the loop-filter gain at $E_S/N_0 = \rho$. Now, from [58], [9], and [97] we know that if the loop-filter gain needed to achieve $\zeta = 0.95$ and $\omega_n = 8.24 \cdot 10^{-3}/T$ in noiseless conditions is $K_{a,DD}(\infty)$, then in order to maintain the same ζ and ω_n at $E_S/N_0 = \chi$ the loop-filter gain $K_{a,DD}(\chi)$ must be $K_{a,DD}(\chi) = K_{a,DD}(\infty)/\alpha_{\chi,DD}$ if AGC effects are ignored by assuming $K=1$ (as done in [58], [9], [97]). Moreover, Sec. 1.5 shows that to assume $K=1$ is unrealistic and AGC effects must be modeled, and it is easily shown that in practice the loop-filter gain at $E_S/N_0 = \chi$ actually must be $K_{a,DD}(\infty) \cdot \beta_{\infty,DD}/\beta_{\chi,DD}$.

Thus, when inspecting Figs. 46-47 it is important to realize that since the loop-filter's gain is different at each SNR, the results for a given phase detector cannot be obtained using a *single* PLL with a *fixed* loop-filter gain. The correct way to interpret Figs. 46-47 (and similar graphs in texts such as [22] and [21]) is hence by considering the results at each SNR as if they were obtained by measurements on PLLs *unique* to *that* SNR that were *optimized for operation* at *that* SNR to yield $\zeta = 0.95$ and $\omega_n = 8.24 \cdot 10^{-3}/T$. Indeed, as noted earlier, a *single* PLL with a fixed loop-filter gain and which uses c_M, DD_M or d_M cannot maintain constant ζ and ω_n over the entire SNR range; rather, the desired values will be attained only at a *single* SNR which we call the *optimization SNR*. When we are not operating at the optimization SNR the PLL will observe changes in ζ and ω_n that will cause changes in all of the PLL's parameters, e.g. the noise bandwidth $B_L = \frac{1}{2}\omega_n(\zeta + 1/(4\zeta))$, the settling time $T_{set} \approx 2\pi/\omega_n$, the lock range $\Delta\omega_L \approx 2\zeta\omega_n$, the pull-out range $\Delta\omega_{PO} \approx 1.8\omega_n(\zeta + 1)$, and the cycle-slip statistics ([103 Chap. 2],[22 Sec. 6.4],[49 Chap. 6], [58], [9], [97]). To quantify this effect, suppose we use a phase detector $P(n)$ in a PLL that has a fixed-gain loop-filter, and that the PLL is designed for operation at $E_S/N_0 = \lambda$ (the "optimization SNR") with the optimal parameters ω_{opt} and ζ_{opt} (e.g., λ might be the lowest SNR at which the error correction decoder provides an acceptable coding gain). It can be shown [48 Sec. 9.1] that, accounting for AGC effects, at $E_S/N_0 = \chi$ the natural frequency $\omega_{n\chi}$ and damping ratio ζ_χ will be:

$$\omega_{n\chi} = \omega_{opt}\sqrt{\beta_{\chi,P}/\beta_{\lambda,P}} \text{ and } \zeta_\chi = \zeta_{opt}\sqrt{\beta_{\chi,P}/\beta_{\lambda,P}}. \tag{89}$$

Thus (assuming $\beta_{\chi,P}$ increases monotonically vs. χ) we conclude from (89) that for $\chi > \lambda$ we have higher-than-optimal ζ and ω_n, and for $\chi < \lambda$ we have lower-than-optimal ζ and ω_n. Only at $\chi = \lambda$ does the PLL perform as desired.

To illustrate this phenomenon and its effect upon the PLL we present in Fig. 55 phase-error variance results and CRBs, computed via (69) and (86). Clearly, as Fig. 55 shows, the variation of ζ and ω_n due to the AGC profoundly affects the PLL. Note that at low SNR the AGC appears to cause a reduction in $\text{var}(\theta_e)$, but it would be fallacy to say that this is a positive effect, since this reduction is due to the reduction of ζ and ω_n, which has a detrimental effect on the PLL's stability, lock range, pull-out range, etc., as outlined in the previous paragraph.

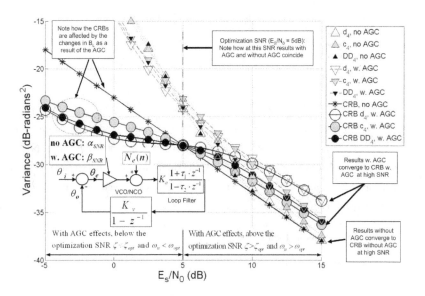

Fig. 55. Calculated $\text{var}(\theta_e)$ using linearized model when AGC effects are ignored and when AGC effects are modeled. Modulation is QPSK (M=4). Optimization SNR is $\lambda = 5\,\text{dB}$. Note that in the linearized model (lower left) the loop noise N_e is computed differently for the "w. AGC" and "no AGC" cases (see Sec. 3.3.2). Also, when AGC effects are ignored, then the loop-filter's gain K_a is selected at <u>each SNR</u> such that $\zeta = \zeta_{opt} = 0.95$ and $\omega_n = \omega_{opt} = 8.24 \cdot 10^{-3}/T$ at <u>each SNR</u>. When AGC effects are modeled, K_a is *constant* and is selected so that at $E_S/N_0 = 5\,\text{dB}$ the PLL has $\zeta = \zeta_{opt} = 0.95$ and $\omega_n = \omega_{opt} = 8.24 \cdot 10^{-3}/T$ (since K_a is *constant*, this will not hold at other SNRs; rather, changes will occur as per (89)).

From (89) it follows that to achieve ω_{opt} and ζ_{opt} at <u>all SNR</u>, the phase detector must have $\beta_{SNR,P} = 1$ (implying a constant gain[14] vs. the SNR since from (66) $g_P(M,K,\chi) = g_P(M,1,\infty) \cdot \beta_{\chi,P}$). We now present such a detector.

3.6   A Family of Robust Adaptive Phase Detectors

In this section we endeavour to present a family of adaptive phase detectors. This adaptive phase detector is composed of two complementing and related structures: (a) the M-PSK phase detector presented in Sec. 3.4, and (b) the M-PSK lock detector presented in Chapter 2. In the following subsection we briefly review some of the derivation of Chapter 2 which are necessary for the current undertaking.

3.6.1   Review of lock detector

In   Chapter   2   an   auxiliary   random   process   was   introduced   via

$$x_M(n) \triangleq \frac{\mathrm{Re}[(I(n)+j \cdot Q(n))^M]}{\left(I^2(n)+Q^2(n)\right)^{\frac{M}{2}}} = \frac{\sum_{k=0}^{M/2} \binom{M}{2k}(-1)^k I^{M-2k}(n)Q^{2k}(n)}{\left(I^2(n)+Q^2(n)\right)^{\frac{M}{2}}} .$$ Elementary rectangular-to-

polar manipulations and the use of De Moivre's theorem [54 eq. 6.9] yielded:

$$x_M(n) = \mathrm{Re}[(I(n)+j \cdot Q(n))^M] \Big/ \left(I^2(n)+Q^2(n)\right)^{\frac{M}{2}} = \cos(M\varphi_n). \tag{90}$$

As clearly seen in (90) and as elaborated upon in Chapter 2, the value of $x_M(n)$ is independent of K and, hence, of the AGC. A lock detector was defined (see Sec. 2.3.1) as $\hat{l}_{M,N} \triangleq \frac{1}{2N} \sum_{n=-N+1}^{N} x_M(n)$. It was found that when the carrier loop is unlocked $E[\hat{l}_{M,N}]=0$. Conversely, when locked, at $E_S / N_0 = \chi$ we have (Sec. 2.4):

---

[14] Sometimes, the optimal value of ω_n will be SNR dependent [48 Chaps. 7, 8]. However, even in that case the desired variation of ω_n will not usually corresponds to the variation induced by the changing phase detector gain, so we would rather have a constant-gain PD and modify ω_n via changing the loop filter coefficients adaptively as a function of the SNR (as estimated by an SNR estimator, see Chapter 4).

$$\tilde{f}_M(\chi) \triangleq E\left[\hat{i}_{M,N} \middle| E_S/N_0 = \chi\right] = \hat{i}_{M,\infty}\middle|_{E_S/N_0=\chi} = E\left[\cos\left(M\Delta\phi_n\right)\right]\cdot E\left[\cos\left(M\theta_e\right)\right]$$
$$= \left(\int_{-\pi}^{\pi}\cos(M\Delta\phi)p_R\left(\Delta\phi\middle|\chi\right)d\Delta\phi\right)E\left[\cos\left(M\theta_e\right)\right].$$
(91)

Due to $x_M(n)$'s independence vis-à-vis the AGC, the same is true for $\hat{i}_{M,N}$. Regarding the distribution of $\hat{i}_{M,N}$, if we ignore the term $E\left[\cos\left(M\theta_e\right)\right]$ (which is a very good approximation, according to the data presented in Sec. 2.4) a good approximation is (see Sec. 2.5):

$$\hat{i}_{M,N}\middle|\ locked \sim N\left(f_M(\chi)\,,\,1/(2N)\right)$$
$$\hat{i}_{M,N}\middle|\ unlocked \sim N\left(0\,,\,1/(2N)\right)$$
(92)

where $f_M(\chi)$ is given in (30)-(32). Further analysis of $\hat{i}_{M,N}$ can be found in Chapter 2.

Using $\hat{i}_{M,N}$ and $d_M(n)$ we now define the adaptive phase detection structure whose analysis is the primary goal of this section.

3.6.2   Definition and structure

The second new phase detector structure presented in this book is an adaptive phase detector that will be shown to have unity gain during tracking, hence allowing for optimal loop parameters to be maintained at all SNRs where the PLL can lock (see discussion in Sec. 3.5). The idea behind this adaptive detector is simple: if we somehow estimate g_d in real-time, and divide $d_M(n)$ by this estimate, we arrive at a constant-gain detector. Fortunately, this is easy to do. To see this, we evoke (92) and the fact that every random variable is an unbiased estimate of its expectation to write:

$$\hat{i}_{M,N} \approx f_M(\chi)$$
(93)

which reveals that, when in lock, $\hat{i}_{M,N} \approx \alpha_{\chi,d} = \beta_{\chi,d} = (1/M)g_d$. Thus, if we define $V_{M,N}(n) \triangleq d_M(n)/(M\cdot\hat{i}_{M,N})$, such a phase detector should have unity gain. When out of lock, $\hat{i}_{M,N} \approx 0$ (see Chapter 2) so an appropriate "worst case" or otherwise defined value μ needs to be substituted for $\hat{i}_{M,N}$ in the expression for $V_{M,N}(n)$ in order to achieve acceptable performance during acquisition (μ is discussed further in Sec. 3.6.6). We hence define the adaptive phase detector as follows:

$$V_{M,N}(n) \triangleq d_M(n)/(M \cdot \delta) \quad \text{with} \quad \delta = \begin{cases} \hat{I}_{M,N} & \text{Carrier is locked} \\ \mu & \text{Carrier is unlocked} \end{cases} \tag{94}$$

A diagram of a fixed-point hardware implementation structure for $V_{M,N}(n)$ is presented in Fig. 56. Observe how the division by $2N$ is avoided, where it is assumed that $2N$ is a power of 2 (see also Sec. 2.3.3).

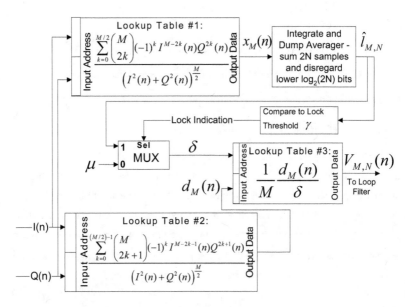

Fig. 56. Hardware implementation of $V_{M,N}(n)$.

3.6.3   Comments regarding the hardware implementation of $V_{M,N}$

Regarding Fig. 56, the implementation of LUT (Lookup Table) #1 was discussed in Sec. 2.3.3, and implementation of LUT #2 was discussed in Sec. 3.4.8. There it was shown that LUTs #1 and #2 can be efficiently realized in fixed-point hardware. As for LUT #3, observe that the lowest value of $\hat{l}_{M,N}$ that need be handled is the lock threshold value, since below this value μ is used. Hence the largest absolute value that LUT #3 needs to accommodate is $(\sup|d_M(n)|)/(M \cdot \min\{\mu,\gamma\}) = 1/(M \cdot \min\{\mu,\gamma\})$, with γ being the lock threshold. Typically, $\mu \geq \gamma$ (see Sec. 3.6.6) and neither parameter would be less than about 0.04, since below that value the SNR is so low that there is scarcely hope of the PLL locking (see Fig. 31). This means that $1/(M \cdot \min\{\mu,\gamma\}) \leq 25/M$, and thus the dynamic range of the data in LUT #3 can be sufficiently limited to allow for its compact fixed-point hardware implementation.

It should be noted that although from an implementational efficiency standpoint the implementation of $V_{M,N}(n)$ can be implemented with the same N as $\hat{l}_{M,N}$ as shown in Fig. 56, this does not have to be the case. This is especially true if the values of N needed to compute $\hat{l}_{M,N}$ in the presence of fading are different from those which are ideal for $V_{M,N}(n)$ (see footnote 15 on page 115, Secs. 2.7 and 3.6.8, and also [104]) .

3.6.4   Linear modeling of $V_{M,N}$

The linear model analysis of $V_{M,N}$ shall now be done by relying on the analysis in Sec. 3.4 regarding d_M .

a)   S-Curve, gain, and amplitude suppression factors
Assuming that the loop is locked, from (76) and (94):

-113-

$$V_{M,N}(n) = \frac{d_M(n)}{M \cdot \hat{l}_{M,N}} = \frac{\sin(M\varphi_n)}{M \cdot \hat{l}_{M,N}}$$

$$= \frac{\cos(M\Delta\phi_n)}{M \cdot \hat{l}_{M,N}}\sin(M\theta_e) + \frac{\sin(M\Delta\phi_n)}{M \cdot \hat{l}_{M,N}}\cos(M\theta_e).$$

(95)

To find the linear-model parameters of $V_{M,N}$, we choose to treat $\hat{l}_{M,N}$ as a constant. This is justified by noting that $\hat{l}_{M,N}$ changes significantly slower than $d_M(n)$ (slower by a factor of $2N$, which typically would be at least in the order of 100 (see Sec. 2.6 and Sec. 3.6.8)). With that assumption and since $E[\sin(M\Delta\phi_n)] = 0$, we have from (95) that the S-Curve is

$$S_V(\theta_e) \triangleq E[V_{M,N}(n)|\theta_e] = E[\cos(M\Delta\phi_n)]\sin(M\theta_e)/(M \cdot \hat{l}_{M,N})$$

$$= (f_M(\chi)/(M \cdot \hat{l}_{M,N})) \cdot \sin(M\theta_e)$$

(96)

and it is easy to show from (96) and (64)-(66) that:

$$g_V(M,K,\chi) = \alpha_{\chi,V} = \beta_{\chi,V} = f_M(\chi)/\hat{l}_{M,N}$$

(97)

The central idea here is, again, that $\hat{l}_{M,N} \approx f_M(\chi)$ so that:

$$g_V(M,K,\chi) = \alpha_{\chi,V} = \beta_{\chi,V} \approx 1.$$

(98)

It is stressed that for all M we have that g_V, $\alpha_{SNR,V}$, and $\beta_{SNR,V}$ approximately equal 1 independent of *(a)* the value of K and the AGC's performance, and *(b)* the SNR.

b)   Loop noise, self noise, and squaring loss of $V_{M,N}(n)$

Once again treating $\hat{l}_{M,N}$ as a constant approximately equal to $f_M(\chi)$, we see that the only random process in (94) is $d_M(n)$. Thus, the squaring loss of $V_{M,N}$ should be identical to that of d_M (shown in Fig. 45). Formally, it is easily shown from (67), (68), (93) and (95)-(96) that $N_{e,V} \approx N_{e,d}/f_M(\chi)$, $\xi_V \approx \xi_d/f_M^2(\chi)$, and (since $\alpha_{\chi,V} = \beta_{\chi,V} \approx 1$) that indeed $\Omega_V \approx \xi_V/1^2 \approx \xi_d/f_M^2(\chi) = \Omega_d$. Hence, from (69) we conclude that d_M and $V_{M,N}$ have the same phase-error variance performance, but with the important difference that results for $V_{M,N}$ can be achieved using a <u>fixed-gain</u> loop-filter (see Sec. 3.5) ; nonlinear-model simulations presented in Sec. 3.6.5 verify this.

Table 3 summarizes the main linearized-model parameters for the phase detectors discussed in this book. Fig. 57 shows plots of α_{SNR} for the phase detectors discussed in this book. Plots of β_{SNR}, assuming the example AGC parameters of Sec. 1.5, are given in Fig. 58. Comparing Fig. 57 to Fig. 58, we can see that the AGC's effect upon DD_M and c_M is quite pronounced. Moreover, since only $V_{M,N}$ has a constant β_{SNR}, only $V_{M,N}$ will be able to maintain optimal loop parameters at all SNRs (see Sec. 3.5).

The reasons for the AGC's effect on $\beta_{SNR,DD}$ and $\beta_{SNR,c}$ (and its particularly striking effect on $\beta_{SNR,c}$) is easily understood by looking at a graph of K and K^M, given in Fig. 59. Since $DD_M(n) \propto K$ and $c_M(n) \propto K^M$ (see Sec. 3.4.8) then, as seen, then AGC has a profound and predictable effect on $\beta_{SNR,DD}$ and $\beta_{SNR,c}$.

As a caveat, we note that the delay $2N \cdot T$ incurred during the computation of $\hat{i}_{M,N}$ must not substantially impact the validity of the approximation $\hat{i}_{M,N} \approx f_M(\chi)$ when that value is used to compute $V_{M,N}(n)$. In Sec. 3.6.8, it is shown that a relatively small N is required[15] to achieve good performance over practical SNRs, so the delay is inconsequential for most systems (particularly where the symbol rate is high compared to the channel fading rate, which is usually a very good assumption if suppressed-carrier coherent M-PSK is the chosen modulation (see [78], [79], [21 p. 250]), and this is a particularly good assumption for geosynchronous microwave satellite links (see [10 Chap. 4])). Nonetheless, this constraint must be taken into account when deciding whether usage of $V_{M,N}$ is appropriate.

---

[15] A further constraint exists for N, namely that the desired lock and false-alarm probabilities are attained (see Sec. 2.6). If this constraint conflicts with those of Sec. 3.6.8, the Integrate-and-Dump module in Fig. 56 can be duplicated, with a different N being used in each module. One module would be used to generate $\hat{i}_{M,N}$ for lock detection (and to drive the "sel" input to the MUX), while the other would be used to drive the "1" input to the MUX.

3.6.5 Nonlinear model simulations for $V_{M,N}$

We have already noted in the previous section that $V_{M,N}$ should have the same phase-error variance characteristics of d_M, as displayed in Fig. 46 and Fig. 47. But, as a crucial difference of $V_{M,N}$ vis-à-vis d_M (and, indeed, vis-à-vis c_M and DD_M), $V_{M,N}$ should be able to achieve the results predicted in those figures, at <u>every</u> SNR, using a <u>single</u> PLL with a <u>fixed</u> loop filter.

To validate the above predictions regarding the performance of $V_{M,N}$, simulations were conducted using the nonlinear equivalent baseband model, assuming the AGC of Sec. 1.5. This is shown in Fig. 60 and Fig. 61, and indeed the results for $V_{M,N}$ agree with those in Fig. 46 and Fig. 47 for d_M (though with the crucial difference that $V_{M,N}$ achieves these results with a <u>fixed-gain</u> loop-filter; see Sec. 3.5). As is evident by comparing Fig. 47 to Fig. 60 and Fig. 61, the AGC has a profound effect upon c_M and DD_M, due to the changes in ω_n and ζ as described in Sec. 3.5. For the same reasons, the CRBs for c_M and DD_M become curved (see Sec. 3.5 and Fig. 55; for clarity those CRBs are omitted from Fig. 60 and Fig. 61).

3.6.6 Unlocked-state operation of $V_{M,N}$

From (94), $V_{M,N}$ will exhibit behaviour identical to that of d_M during acquisition, as it is simply $d_M(n)$ multiplied by the constant $1/(M \cdot \mu)$. The gain of $V_{M,N}$ is then $g_V(M,K,\chi) = g_d(M,K,\chi)/(M \cdot \mu) = f_M(\chi)/\mu$. To maintain the optimal loop parameters during acquisition, we strive to have $g_V = 1$, implying that we should aspire for $\mu = f_M(\chi)$. An algorithm for deciding upon an appropriate μ would try, for example, to determine the latter either by: (a) using a worst case value (i.e. the value of $f_M(\chi)$ for the lowest SNR for which operation is desired), (b) using the last measured value of $\hat{l}_{M,N}$ when the receiver was locked (because $E[\hat{l}_{M,N}] = f_M(\chi)$), or (c) using some SNR estimation technique (such as measurements on an auxiliary pilot signal) to indirectly estimate $f_M(\chi)$. It is important to note that performance during acquisition is only partially addressed by using a constant μ, due to the fact that, since $f_M(\chi)$ varies with the SNR yet μ is constant, $g_V = f_M(\chi)/\mu$ will vary vs. the SNR.

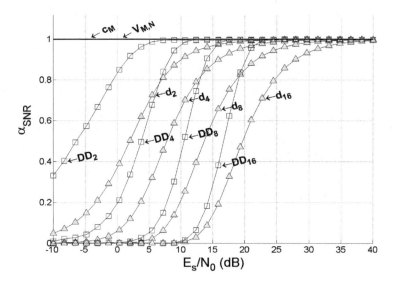

Fig. 57. α_{SNR} for the various phase detectors discussed in this book.

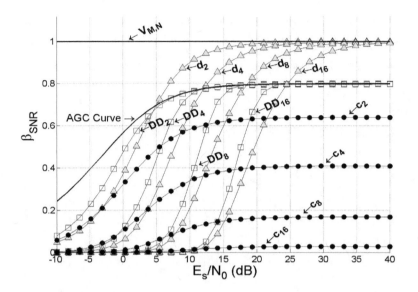

Fig. 58. β_{SNR} for the various detectors discussed in this book, assuming the AGC that is described in Section 1.5. Also plotted is the AGC curve (i.e. $K = \Upsilon_{AGC}(E_S/N_0)$).

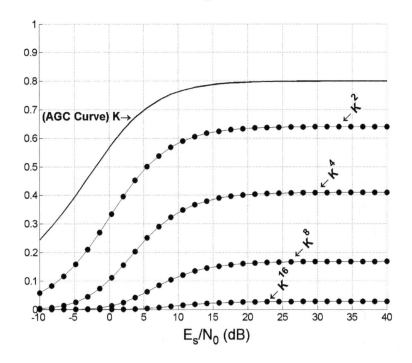

Fig. 59. K and K^M as a function of the SNR for the AGC of Section 1.5.

Table 3. Comparison of important linearized-model phase detector characteristics

PD	Self Noise $\xi = 2\cdot\chi\cdot\mathrm{var}(N_e)$	Linearized Gain g_L	α_{SNR} and β_{SNR}
c_M	$\xi_c = K^{2M}\cdot\dfrac{\left[(M-1)!\sum\limits_{i=0}^{M-1}\binom{M}{i}\dfrac{\chi^i}{i!}\right]}{\left(M\cdot\chi^{M-1}\right)}$ $\underbrace{\qquad\qquad}$ from eq. (17) in [92]	$g_c = M\cdot K^M$ Source: [22 Eq. 6-116]	$\alpha_{\chi,c}=1$ $\beta_{\chi,c}=K^M$
DD_M	$\xi_{DD} = 2\cdot\chi\cdot K^2\cdot\mathrm{var}(N_{e,DD}\,\vert\,K=1)$ $\underbrace{\qquad\qquad}$ from eq. (3) in [58]	$g_{DD}=K\cdot\alpha_{\chi,DD}$ Source: [9]	$\alpha_{\chi,DD}$ from eq. (10),(11),(32) in [9] $\beta_{\chi,DD}=K\cdot\alpha_{\chi,DD}$
d_M	$\xi_d = \dfrac{\chi}{M^2}\left(1-\dfrac{\sqrt{\pi\chi}}{2}\cdot e^{-\chi/2}\cdot\left[I_{(2M-1)/2}\left(\tfrac{\chi}{2}\right)+I_{(2M+1)/2}\left(\tfrac{\chi}{2}\right)\right]\right)$	$g_d = M\dfrac{\sqrt{\pi\chi}}{2}\cdot e^{-\chi/2}$ $\times\left[I_{\frac{M-1}{2}}\left(\tfrac{\chi}{2}\right)+I_{\frac{M+1}{2}}\left(\tfrac{\chi}{2}\right)\right]$	$\beta_{\chi,d}=\alpha_{\chi,d}$ $=\dfrac{\sqrt{\pi\chi}}{2}e^{-\chi/2}\left[I_{\frac{M-1}{2}}\left(\tfrac{\chi}{2}\right)+I_{\frac{M+1}{2}}\left(\tfrac{\chi}{2}\right)\right]$
$V_{M,N}$	$\xi_V \approx \dfrac{\dfrac{\chi}{M^2}\left(1-\dfrac{\sqrt{\pi\chi}}{2}\cdot e^{-\chi/2}\cdot\left[I_{(2M-1)/2}\left(\tfrac{\chi}{2}\right)+I_{(2M+1)/2}\left(\tfrac{\chi}{2}\right)\right]\right)}{\left(\dfrac{\sqrt{\pi\chi}}{2}e^{-\chi/2}\cdot\left[I_{(M-1)/2}\left(\tfrac{\chi}{2}\right)+I_{(M+1)/2}\left(\tfrac{\chi}{2}\right)\right]\right)^2}$	$g_V \approx 1$	$\alpha_{\chi,V}\approx1$ $\beta_{\chi,V}\approx1$

Notes: (a) To ignore AGC effects, substitute $K=1$; (b) Results for $V_{M,N}$ assume $\hat{I}_{M,N}\approx f_M(\chi)$ (see Sec 3.6.8) (c) While $\xi_V\geq\xi_d$, this <u>does not</u> result in an increase in $\mathrm{var}(\theta_e)$, as substitution of the appropriate variables into (69) shows; (d) The expressions involving Bessel functions can be simplified further into finite sums of terms which include only exponents and polynomials (see Appendix A).

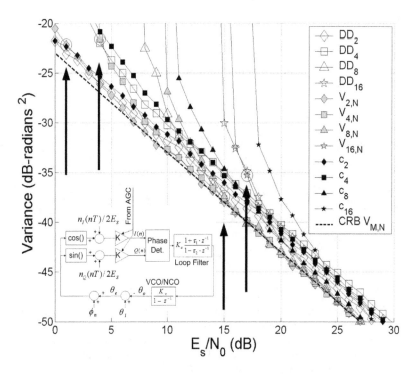

Fig. 60. $\mathrm{var}(\theta_e)$, obtained via nonlinear-model simulations including AGC effects. The loops for DD_M and c_M are optimized for input SNRs of 1, 4, 15, and 17 dB, for $M = 2$, 4, 8, and 16, respectively, where at those SNRs we desire to have $\zeta = 0.8$ and $\omega_n = 0.011/T$. For $V_{M,N}$, $\zeta = 0.8$ and $\omega_n = 0.011/T$ throughout. The arrows aid in finding the results for $V_{M,N}$ at the optimization SNRs (those data points are also circled). For $V_{M,N}$, $N = 256$. Note that DD_M appears to outperform $V_{M,N}$ at low SNR; but this is a fallacy, since this apparent advantage is due to the reduction of ω_n, ζ, and B_L in the PLLs that use DD_M. See Section 3.6.7.

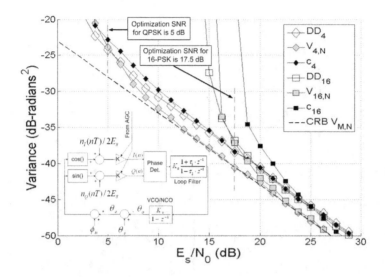

Fig. 61. var(θ_e) for QPSK and 16-PSK obtained via nonlinear-model simulations including AGC effects. K behaves according to the AGC of Sec. 1.5. For $V_{M,N}$, $N=256$ was chosen. K_a is *constant* and is selected so that at the optimization SNRs (which are 5 dB and 17.5 dB for QPSK and 16-PSK, respectively) the PLLs have $\zeta = \zeta_{opt} = 0.95$ and $\omega_n = \omega_{opt} = 8.24 \cdot 10^{-3}/T$. Since K_a is constant while $\beta_{SNR,DD}$ and $\beta_{SNR,c}$ are variable (See Fig. 58), then $\zeta = \zeta_{opt}$ and $\omega_n = \omega_{opt}$ is not true at other SNRs for the loops employing DD_M and c_M; rather, changes will occur as per (89). However, since $\beta_{SNR,V} \approx 1$ at all SNR, for the PLLs employing $V_{M,N}$ we do have $\zeta = \zeta_{opt}$ and $\omega_n = \omega_{opt}$ at all SNR. Note that DD_M appears to outperform $V_{M,N}$ at low SNR; but this is a fallacy, since this apparent advantage is due to the reduction of ω_n and ζ in the PLLs that use DD_M. See Sec. 3.6.7 and Fig. 63.

3.6.7   System identification analysis

To get a qualitative feel of the operation of $V_{M,N}$, Fig. 62 compares the step response for carrier PLLs which use DD_M, c_M, and $V_{M,N}$. The upper subplot of that figure shows the output phase trajectory for a single input data set for each SNR. Because of the input noise, it may be difficult to adequately distinguish the system response from a single data set, especially[16] for the lower SNRs. This difficulty is overcome by using several data sets to drive the systems at each SNR, and for each SNR the measured responses are then averaged. It then becomes easy to discern the systems' responses. This is shown in the bottom subplot of Fig. 62, where it is seen that the system responses using $V_{M,N}$ are virtually identical at all SNRs, while the responses obtained by using c_M and DD_M are strongly dependent upon the SNR.

---

[16] In the top subplot of Fig. 62 it appears that the response for $V_{M,N}(n)$ at $E_S/N_0 = 0$ dB is much noisier than that of the other detectors; but this is because the loop bandwidth for c_M and DD_M decreases at low SNR (see Fig. 63). This reduction in ω_n may have a seemingly positive effect on the phase-error variance, but it has the negative effects of, for example, reducing the lock range, a higher settling time, and a smaller pull-out range (see Sec. 3.5).

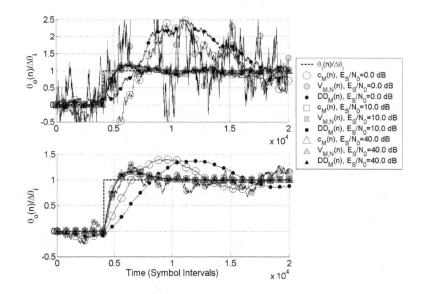

Fig. 62. Simulated responses of carrier PLLs to a phase step of $\theta_i(n)=\Delta\theta_i \cdot u(nT-t_0)$, where $u(t)$ is the unit step function and $t_0 = 4100 \cdot T$. Modulation is QPSK ($M=4$). Loops are optimized for $E_S / N_0 = 10 \, \text{dB}$, where at that SNR we desire $\zeta = 0.8$ and $\omega_n = 9 \cdot 10^{-4}/T$. Upper subplot is the response obtained from a single data set. Bottom subplot is the average of the responses obtained from 100 data sets. For $V_{M,N}$, $N = 2048$. We assume K behaves according to the AGC described in Section 1.5.

We can also arrive at quantitative results by using the Steiglitz-McBride [105] system identification algorithm. To do this, at each SNR we average the PLLs' responses for a sufficient number of input data sets, that is, until the averaged response curves are sufficiently noise-free (like we did in order to arrive at the bottom plot of Fig. 62). Then we can use the Steiglitz-McBride algorithm upon the smoothed response in order to estimate ω_n and ζ. The results attained by following such a procedure are shown in Fig. 63. As was predicted in Sec. 3.6.2, $V_{M,N}$ provides the desired ω_n and ζ over the entire input SNR range; DD_M and c_M do not, and the parameters of their PLLs change according to (89). The strong variations in ω_n and ζ for DD_M and c_M cause a corresponding variation in the PLLs' other parameters (see Sec. 3.5) which means that those PLLs behave very non-optimally when not operating at the optimization SNR. In contrast, we deduce from Fig. 63 that PLLs employing $V_{M,N}$ will maintain optimality at all SNRs.

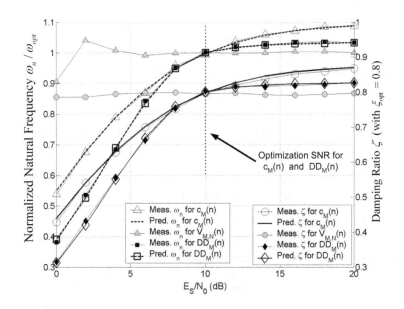

Fig. 63. Predicted and measured performance of DD_M, c_M, and $V_{M,N}$. Modulation is QPSK ($M=4$). For $V_{M,N}$, $N=2048$ was used. PLLs were designed to give $\zeta_{opt}=0.8$ and $\omega_{opt}=9\cdot10^{-4}/T$ at $\chi=10$ dB. We assume that K behaves according to the example AGC described in Section 1.5.

3.6.8 Bounds on N to ensure satisfactory tolerances in PLL parameters in PLLs using $V_{M,N}(n)$

In this section we investigate the parameter N of $V_{M,N}$. In Sec. 3.5 we determined that to maintain optimal PLL parameters we desire $\beta_{SNR,V}=1$ identically, which, since (from (97)) $\beta_{\chi,V}=f_M(\chi)/\hat{I}_{M,N}$, means that we strive to maintain $\hat{I}_{M,N}=f_M(\chi)$. Since (see (92)) $E[\hat{I}_{M,N}]=f_M(\chi)$, the way to achieve acceptable accuracy of the approximation $\hat{I}_{M,N}\approx f_M(\chi)$ is by ensuring that $\hat{I}_{M,N}$'s variance is low enough, which (given (92)) means choosing a high enough N. To obtain a quantitative measure of the value of N that is needed in order to achieve acceptable performance, let us denote the natural frequency and damping factor we are trying to achieve as ω_{opt} and ζ_{opt}. We want to achieve them at **all** $E_S/N_0 = \chi$ in the range $\chi \in [\Gamma, \infty]$ where Γ is some reasonable lower bound (e.g., the PLL's lock threshold). The question is: what is the lower bound on N necessary to ensure, at each $E_S/N_0 = \chi \in [\Gamma, \infty]$, that:

$$P\left(\left|\omega_{n\chi}/\omega_{opt} - 1\right| < tol\right) > C \text{ and } P\left(\left|\zeta_\chi/\zeta_{opt} - 1\right| < tol\right) > C \qquad (99)$$

where tol is the acceptable tolerance for ω_n and ζ, and C is the confidence. Since we want ω_{opt} and ζ_{opt} be achieved for **all** $E_S/N_0 \in [\Gamma, \infty]$, we can define the optimization SNR arbitrarily (and conveniently) as $\lambda = \infty$, and since $\beta_{\infty,V} = 1$ we have from (89) and (97) that

$\omega_{n\chi} = \omega_{opt}\sqrt{f_M(\chi)/\hat{I}_{M,N}}$ and $\zeta_\chi = \zeta_{opt}\sqrt{f_M(\chi)/\hat{I}_{M,N}}$. Straightforward manipulations then show that an equivalent constraint to (99) is:

$$P\left(f_M(\chi)\left((1+tol)^{-2} - 1\right) < \hat{I}_{M,N} - f_M(\chi) < f_M(\chi)\left((1-tol)^{-2} - 1\right)\right) > C. \qquad (100)$$

Since (see (30)) $E\left[\hat{I}_{M,N} | E_S/N_0 = \chi\right] = f_M(\chi)$, to guarantee (100) it suffices that:

$$P\left(\left|\hat{I}_{M,N} - E\left[\hat{I}_{M,N} | E_S/N_0 = \chi\right]\right| < f_M(\chi) \cdot y\right) > C \qquad (101)$$

where $y \triangleq \min\left\{\left|(1-tol)^{-2} - 1\right|, \left|(1+tol)^{-2} - 1\right|\right\}$. Since $\hat{I}_{M,N}$ is Gaussian (see (92)) then (101) is equivalent to

$$erf\left(f_M(\chi) \cdot y \Big/ \left(\sqrt{2}\sqrt{\operatorname{var}\left(\hat{I}_{M,N}\right)}\right)\right) > C \qquad (102)$$

with $erf(x) \triangleq \dfrac{2}{\sqrt{\pi}}\int_0^x e^{-t^2} dt$. Now, since (see (92)) we have $\operatorname{var}(\hat{I}_{M,N}) \leq 1/(2N)$, we can solve (102) for N, whereupon we find that for all $E_S/N_0 = \chi \geq \Gamma$ a suitable lower bound on N would be:

$$N > \left(1/f_M^2(\chi)\right) \cdot \left(erf^{-1}(C)/y\right)^2 \qquad (103)$$

Graphs for N, computed in this manner, are shown in Fig. 64. From that figure we see that, for example, $N=256$ is sufficient to ensure $tol=20\%$ and $C=85\%$ for SNRs above -2, 4, 10, and 16 dB for M=2, 4, 8, and 16. $N=1024$ is sufficient to ensure $tol=20\%$ and $C=85\%$ for SNRs above -5, 2, 8, and 14 dB for M=2, 4, 8, and 16. Thus, only a relatively small N is needed to guarantee optimal PLL parameters above reasonable lock thresholds [58 Sec. III] for the respective modulations.

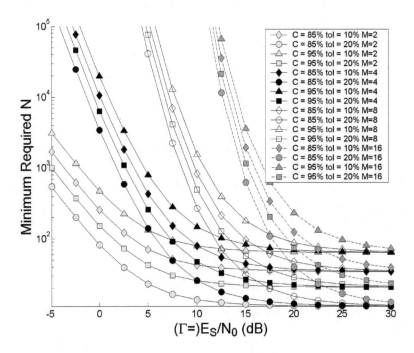

Fig. 64. Lower bound on N needed to achieve ω_{opt} and ζ_{opt} to a desired tolerance, at a given confidence.

3.7 Conclusions

In this chapter we presented and investigated two new families of M-PSK NDA carrier phase detectors for operation in carrier synchronization PLLs in feedback-topology M-PSK coherent receivers.

First, a new family of self-normalizing phase detectors was proposed, and its properties analyzed. It was found that the suggested detectors could be of substantial practical significance, as they have good self-noise performance, have phase-variance

- 128 -

performance which is better than M^{th}-order nonlinearity detectors and comparable to DD detectors (according to simulations), and lend themselves to simple hardware implementation, as a compact, fixed-point lookup table. They also possess self-normalizing qualities that simplify the receiver design by significantly decoupling the AGC circuit from the carrier synchronization PLL.

Next, we investigated an adaptive phase detection structure for M-PSK. This detector structure was characterized via theoretical derivations, simulation results, and system identification analysis. The major novelty of this family of adaptive phase detectors is that it has a constant gain during tracking, which, as was shown, allows a PLL that uses the proposed structure to maintain optimal PLL parameters at any SNR at which the PLL can attain lock. It is emphasized that these optimal parameters are maintained even though the PLL has a fixed (i.e. non-adaptive) loop filter. As an additional advantage, the detector was found to be inherently independent of the AGC's operating point and performance. Moreover, theoretical derivations using the linear model as well as simulations using the nonlinear model have shown that the detector has superior phase-error variance performance, as compared to DD detectors and M^{th}-order nonlinearity detectors.

Both families of detectors were shown to have a compact fixed-point hardware implementation that is suitable for use within an FPGA or ASIC. Thus, they have immediate applications in contemporary coherent M-PSK receivers.

Chapter 4   New Methods for Real-Time Generation of SNR Estimates for Digital Phase Modulation Signals

Part A   A New SNR Estimation Structure for M-PSK

4.1   Introduction

In any M-PSK receiver, one of the most important metrics that can be generated is an estimate of the channel E_S / N_0 ratio. In many modern communications schemes an accurate E_S / N_0 estimate is needed not only as a monitoring aid, but it plays an important role in the receiver's operation. For example, some error correction decoders can make use of an E_S / N_0 estimate to increase their coding gain (e.g. turbo codes [12]). Another example are systems that employ diversity reception [13 Sec. 14.4], for which SNR estimates are used to assign relative weights to the data obtained from the various receivers. Yet another example are adaptive schemes where the data and/or coding rates are altered according to the E_S / N_0 (e.g. [14], [15]). The reader is referred to [16 Sec. 1.2] and the references therein for an extensive overview of these and other applications of SNR estimates in communications systems.

In this chapter we shall present a quantitative analysis of the SNR estimation method suggested in Chapter 2, as outlined in Section 2.3.5. A focus of this analysis will be around the observation that the need for E_S / N_0 estimates is not fulfilled merely by facilitating their availability; the estimates must also be timely. In this respect, the method analyzed here is shown to produce accurate estimates using only a small number of symbols, thus facilitating the generation of a rapidly updating estimate. We shall

---

This chapter was published in part in Linn, Y., "Quantitative Analysis of a New Method for Real-Time Generation of SNR Estimates for Digital Phase Modulation Signals", *IEEE Transactions on Wireless Communications*, vol. 3, no. 6, pp. 1984-1988, Nov. 2004, and in Y. Linn, "A Real-Time SNR Estimator for D-MPSK over Frequency-Flat Slow Fading AWGN Channels," in *Proc. 2006 IEEE Sarnoff Symposium*, Princeton, NJ, Mar. 27-28, 2006. Parts will also appear in "A Carrier-Independent Non-Data-Aided Real-Time SNR Estimator for M-PSK and D-MPSK Suitable for FPGAs and ASICs", IEEE *Transactions on Circuits and Systems I*, in press.

furthermore show that other advantages of the proposed method is that it is Non Data Aided, operates at a rate of one sample per symbol (which corresponds to the symbol strobe), and has a simple implementation that is easily realizable in an FPGA or ASIC.

Since the SNR estimation method presented in Chapter 2 will only work if the carrier PLL is locked, this is assumed to hold for this part of the current chapter (Part A). However, in Part B of this chapter we dispense with that assumption and present an SNR estimator that also works in the absence of carrier synchronization.

There have been many SNR estimation algorithms proposed by various researchers in the past. For example, the reader is referred to [59], [60], [61], [62], [63], [64], [65], [66], [67], [74]. We shall not address all of those estimators individually, since this would take an inordinate amount of space, and, moreover, as we shall show that this is unnecessary. In lieu of that, we shall conduct our quantitative comparison versus some of the most widely used SNR estimation methods, namely, (a) SNR estimation from the SER (Symbol Error Rate) [62]; (b) the M_2M_4 estimator [60]; and (c) the SVR estimator [60]. We shall supplement this quantitative comparison with qualitative comparisons versus other SNR estimators that will show that the estimator proposed here possesses several important advantages over these previously proposed estimators. Of those advantages, the most endearing one is the fact that the proposed estimator has an exceptionally simple hardware structure which is almost trivial to implement with FPGAs or ASICs. Moreover, the fact that the estimator is NDA (Non Data Aided) and requires only sample per symbol sets it apart from most other estimators previously cited.

4.1.1 The general principle behind SNR estimation

In general, there are two SNR estimator types: Data-Aided (DA) and Blind (or Non Data Aided (NDA)). Data Aided methods use known symbols that are embedded in the data stream in order to estimate the SNR. For example, we could estimate the SNR from measuring the error rate on a preamble or pilot sequence. Non Data Aided methods use a nonlinearity upon the received signal to generate an SNR-dependent metric, which is then used to estimate the SNR. For both types of detectors, the SNR estimation principle is as follows. We want to estimate the E_s / N_0 ratio, which we denote in this book using the symbol χ. We denote the SNR estimate as γ. In general terms, we first compute an observation variable ℓ based upon many symbols of the input signal. The idea is to choose a computation process that will yield an ℓ for which $E[\ell \mid E_S / N_0 = \chi]$ is a strictly monotonic function of χ that we shall call $f(\chi)$ (i.e. $f(\chi) \triangleq E[\ell \mid E_S / N_0 = \chi]$). Then, we can compute an estimate of χ via $\gamma = f^{-1}(\ell)$. This process is shown in Fig. 65.

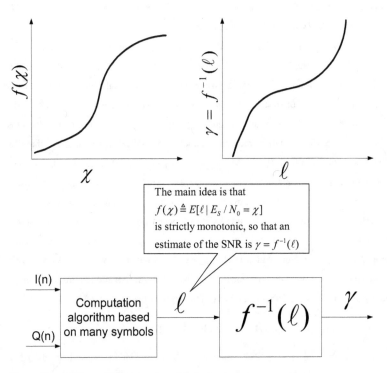

Fig. 65. SNR estimation principle.

Specifically, for our SNR estimator discussed in Part A of this chapter, we choose $\ell = \hat{i}_{M,N}$ (defined in Sec. 2.3.1) and $f(\chi) = f_M(\chi)\left(\triangleq E\left[\hat{i}_{M,N} \mid E_S / N_0 = \chi\right]\right)$ (defined in Sec. 2.3.2). We shall now proceed to formally define this process and evaluate it quantitatively and qualitatively.

4.2 Review of Receiver structure and Lock Detector

4.2.1 Signal and receiver models

The signal and receiver models, as well as the applicable notations, have been defined in Section 1.4. We shall at first assume that the SNR is constant, i.e. we ignore the effects of possible fading; these shall be treated in Sec. 4.7.

4.2.2 Lock detector definition, expectation and distribution

Here, we review the necessary derivations made in Chapter 2. Since this is only a short review of some results, the reader is urged to re-examine that chapter for more detailed information.

First, we define a process:

$$
x_{M,n} \triangleq \frac{\mathrm{Re}[(I(n)+j\cdot Q(n))^M]}{\left(I^2(n)+Q^2(n)\right)^{\frac{M}{2}}} = \frac{\sum_{k=0}^{M/2}\binom{M}{2k}(-1)^k I^{M-2k}(n)Q^{2k}(n)}{\left(I^2(n)+Q^2(n)\right)^{\frac{M}{2}}}
\tag{104}
$$

A new type of lock detector was defined in Chapter 2 through:

$$
\hat{l}_{M,N} = \frac{1}{2N}\sum_{n=-N+1}^{N} x_{M,n}
\tag{105}
$$

When the carrier loop is unlocked, it was shown in Section 2.3.2 that $E[\hat{l}_{M,N}]=0$. Conversely, when locked, we have (from eq. (44)) :

$$
\begin{aligned}
\tilde{f}_M(\chi) &= \left[E\left[\cos\left(M\Delta\phi_n\right)\right]\cdot E\left[\cos\left(M\theta_e\right)\right]\right]_{E_S/N_0=\chi}\\
&= f_M(\chi)E\left[\cos\left(M\theta_e\right)\middle| E_S/N_0=\chi\right]
\end{aligned}
\tag{106}
$$

where $\Delta\phi_n \in [-\pi,\pi]$ has the Rician phase distribution (see (29)):

$$
\begin{aligned}
p_R\left(\Delta\phi\middle|\chi\right) &\triangleq p\left(\Delta\phi_n = \Delta\phi\middle| E_S/N_0 = \chi\right)\\
&= \frac{\exp(-\chi)}{2\pi}\times\left[1+\sqrt{2\chi}\cos(\Delta\phi)\exp\left(\chi\cdot\cos^2(\Delta\phi)\right)\cdot\int_{-\infty}^{\cos(\Delta\phi)\sqrt{2\chi}} e^{-y^2/2}dy\right]
\end{aligned}
\tag{107}
$$

From the central limit theorem on eq. (105), $\hat{l}_{M,N}$ has a Gaussian distribution, and, furthermore:

$$\text{var}\left(\hat{l}_{M,N}\right) = \left(\frac{1}{2N}\right) \cdot \left(E[x_{M,n}{}^2] - \left(E[x_{M,n}]\right)^2\right) \tag{108}$$

An important bound on the variance of $\hat{l}_{M,N}$ is also recalled from (49):

$$\text{var}\left(\hat{l}_{M,N}\right) \leq \frac{1}{2N} \tag{109}$$

Note that eq. (109) is valid for both locked and unlocked states, and for any phase error jitter conditions.

4.2.3   Low jitter approximations

Neglecting the effects of carrier phase jitter, (106) becomes:

$$\tilde{f}_M(\chi) = f_M(\chi) = \int_{-\pi}^{\pi} \cos\left(M\Delta\phi\right) \cdot p_R\left(\Delta\phi\middle|\chi\right) \cdot d\Delta\phi \tag{110}$$

where closed-form expressions for (110) are given in (31)-(32) and Table 1. In the case where carrier phase jitter can be neglected we can also deduce from (108) that:

$$
\begin{aligned}
\text{var}\left(\hat{l}_{M,N}\right) &= \left(\frac{1}{2N}\right) \cdot \left(\int_{-\pi}^{\pi} \cos^2\left(M\Delta\phi\right) \cdot p_R\left(\Delta\phi\middle|\chi\right) \cdot d\Delta\phi - \left(f_M(\chi)\right)^2\right) \\
&= \left(\frac{1}{2N}\right) \left(\int_{-\pi}^{\pi} \left(\tfrac{1}{2}\cos\left(2M\Delta\phi\right) + \tfrac{1}{2}\right) p_R\left(\Delta\phi\middle|\chi\right) \cdot d\Delta\phi - \left(f_M(\chi)\right)^2\right) \\
&= \left(\frac{1}{2N}\right) \left(\tfrac{1}{2} + \tfrac{1}{2}f_{2M}(\chi) - \left(f_M(\chi)\right)^2\right)
\end{aligned}
\tag{111}
$$

and further closed-form simplifications of (111) are possible using (31)-(32).

4.3   Principle of SNR Estimation from $\hat{l}_{M,N}$

4.3.1   Theoretical basis

$f_M(\chi)$ is a monotonically increasing function and is thus invertible. It is this inverse relation:

$$\gamma = f_M{}^{-1}(\hat{l}_{M,N}) \tag{112}$$

namely the estimation of the E_S / N_0 ratio from the value of $\hat{i}_{M,N}$, which interests us here. The E_S / N_0 estimate is usually desired in units of dB, as follows:

$$\gamma_{dB} = 10 \log_{10} \left(f_M^{-1}(\hat{i}_{M,N}) \right) \tag{113}$$

The same reasoning applies to the case when jitter is modeled, i.e. when we use $\tilde{f}_M(\chi)$ to estimate the lock metric's value. We can then write:

$$\gamma_{dB} = 10 \cdot \log_{10} \left(\tilde{f}_M^{-1}(\hat{i}_{M,N}) \right) \tag{114}$$

Fig. 66 shows simulated and predicted curves of (113) and (114). This figure was generated by first plotting $f_M(\chi) = E\left[\hat{i}_{M,N} \middle| E_S / N_0 = \chi \right]$ vs. $(E_S / N_0)_{dB}$, using the theoretical prediction of (110), as well as closed-loop simulations of $2N = 20000$ symbol intervals in which (105) was computed and (106) was approximated. Then, the graph was reflected through its $y = x$ diagonal to produce the inverse relations, as given in (113) and (114). Results are presented for 2nd order PLL synchronizers with $2B_L \cdot T = 0.01$ and $2B_L \cdot T = 0.1$, where B_L is the PLL's noise bandwidth ($B_L = 0.5\omega_n \left(\zeta + 1/(4\zeta) \right)$ where ω_n is the natural radian frequency of the PLL and ζ is its damping factor.

More specifically, in Fig. 66 the solid lines are plots of the theoretical jitter-free case of $f_M^{-1}(\bullet)$ (i.e. $f_M(\bullet)$ given by eq. (110)). The blank polygons in that figure were obtained in the closed-loop simulations using eq. (105), where a normalized PLL noise bandwidth of $2B_L \cdot T = 0.01$ was employed. The gray polygons are values of $\tilde{f}_M^{-1}(\bullet)$ predicted by (106) with $2B_L \cdot T = 0.01$ and using the time average $\overline{\cos(M\theta_e)}$ to approximate $E[\cos(M\theta_e)]$, where θ_e was measured in the aforementioned simulations. Completing Fig. 66 are curves obtained with a normalized PLL noise bandwidth of $2B_L \cdot T = 0.1$, where the dashed lines were obtained in closed loop simulations using eq. (105), and the black polygons are values of $\tilde{f}_M^{-1}(\bullet)$ predicted using eq. (106), with $\overline{\cos(M\theta_e)}$ approximating $E[\cos(M\theta_e)]$.

In carrier synchronization PLLs $2B_L \cdot T$ is rarely ([58], [22 Chaps. 5,6], [88]) larger than the order of magnitude of $2B_L \cdot T = 0.01$, and is virtually never as high as $2B_L \cdot T = 0.1$. Thus, it can be safely said (in lieu of Fig. 66) that it is permissible to always use (110)

- 135 -

and not bother with trying to predict (106), and that hence SNR estimation can be done using (113) and there is no need to try and predict $\tilde{f}_M^{-1}(\bullet)$ and use (114); this will henceforth be assumed. The case is less clear for using (111) when non-negligible carrier phase error jitter is present; however, since (109) holds for any phase jitter conditions, it is easy to arrive at "worst case" bounds (i.e. assuming $\mathrm{var}\left(\hat{l}_{M,N}\right) = \dfrac{1}{2N}$), which will also be given.

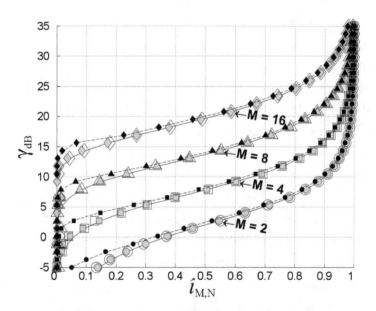

Fig. 66. Simulated and predicted values for eq. (113) and (114).

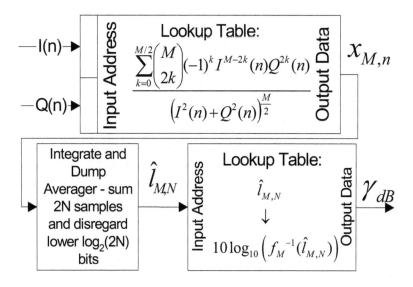

Fig. 67. Efficient hardware generation of γ_{dB}

4.3.2 Hardware implementation

Fig. 67 presents a structure for the generation of (113) in hardware. Since (see Sec. 2.3.3) $|x_{M,n}| \leq 1$, $|\hat{i}_{M,N}| \leq 1$, and since a small dynamic range is needed for (113) (see Secs. 2.3.3 and 2.3.4), the lookup tables can be realized as small fixed-point lookup tables, thus permitting efficient implementation of within an FPGA or ASIC. Note how outright division by $2N$ is avoided in Fig. 67, where for this to be accurate N should be chosen to be a power of 2 (see Sec. 2.3.3). Note also that if $\hat{i}_{M,N}$ is already generated for lock detection then E_S / N_0 estimation requires merely the addition of the small lookup table implementing (113), a trivial addendum.

As a final point for this subsection, we note that a very useful approximation to (113), for "manual" use during the design process (e.g., for designing a rough draft of the receiver), is given by $\gamma_{dB} \approx 10 \log_{10} \left(-M^2 / \left(4 \cdot \ln \left(\hat{i}_{M,N} \right) \right) \right)$ (see (38)). However it must be noted that for accurate SNR estimation in the actual receiver this approximation

- 137 -

should not be used, and, rather, the contents of the lookup table computing γ_{dB} should be computed via numerically evaluating f_M^{-1} by numerically inverting[17] (31)-(32).

4.4   Discussion: Quantitative Measurement of Estimator Performance

We are now interested in obtaining a quantitative evaluation of the efficacy of the proposed estimator.

There are various ways by which this can be done. Some researchers have used the Cramér-Rao Bound (CRB) of the Normalized Mean-Square-Error (NMSE) as a limit to which the NMSE of the SNR estimate is compared (see for example [60], [65]). While this does provide a quantitative measure of the estimate's performance, this benchmark is not as easy to translate into actual estimator design parameters as another performance metric (presented in the next paragraphs). Nonetheless, the NMSE metric is useful because other authors have published results which use this metric, hence facilitating comparison with other estimators (in particular, vs. the data presented in [60]). Hence, we shall present NMSE results in Sec. 4.6.2.

A more useful metric, in the author's opinion, is the metric proposed in [62], which is the following:

How many symbol intervals are necessary in order to generate an SNR estimate to within a desired tolerance, with a desired confidence?

---

[17] Numerically inverting a function $f(x)$ for which the inverse does not have a closed-form representation is a topic that has been studied extensively in the literature. The essential process is (a) evaluate $f(x)$ over a fine enough grid x_i, $i \in \{1, 2, ..., N\}$ in the domain $[x_A, x_B]$ to yield the values $f(x_i) = y_i \in [y_A, y_B]$, $i \in \{1, 2, ..., N\}$, and then (b) determine $x = f^{-1}(y)$ for any $y \in [y_A, y_B]$ by computing the function $f^{-1}(\bullet)$ at y through interpolation from the known coordinate pairs (x_i, y_i), for example using the Lagrange polynomial method or spline curves. For more information see for example [106 Chap. 9-12]. Note that the inverse function (quantized over a fine enough grid) can be stored in a lookup table, so that this process can be done beforehand and not in real-time.

Let us qualify this question mathematically. Suppose that ρ_ℓ is an SNR estimate which is based upon the signal information in the preceding ℓ symbol intervals. The question is as follows: *What is the minimal value of ℓ needed so that the following holds?*

$$P\left(\left|\rho_\ell - \chi\right| < tol\right) > C \tag{115}$$

where in (115) *tol* is the tolerance and C is the confidence (for example, reasonable values would be *tol* $=1$ dB and C=99%).

The answer to such a question can be easily translated into practical conclusions which are pertinent to the design process. To see this, suppose that the value of ℓ that satisfies (115) is ℓ_0. Then this means that in order to achieve an estimate that is accurate to within *tol* with a confidence of C, the designer knows that he/she must ensure that the estimator operate over ℓ_0 symbol intervals, and that the delay incurred for estimation would be $\ell_0 \cdot T$. Conversely, if the maximum allowable estimation delay is Δ_e, then we can calculate the number of symbol intervals allowable for estimation via

$$\ell_0 = \left\lfloor \frac{\Delta_e}{T} \right\rfloor$$ (where $\lfloor \bullet \rfloor$ means "round down to the nearest integer"). This, in turn, can be used to ascertain whether the desired tolerance and confidence requirements can be met by the design.

We shall use the evaluation method proposed in [62] as the performance metric by which we measure the performance of our estimator to estimation via the SER, and this shall be the topic of Sec. 4.5. As noted earlier, for completeness in Sec. 4.6 we shall present comparisons vs. other SNR estimators using the NMSE criterion.

4.5 Comparison of Estimation via $\hat{l}_{M,N}$ to Estimation via the SER

In this section, following the evaluation method proposed in [62], we shall arrive at quantitative results that describe the minimum amount of symbols necessary to arrive at an SNR estimate from $\hat{l}_{M,N}$ to within a desired tolerance, with a desired confidence. These results will be compared to the number of symbols necessary to estimate the SNR from the SER, hence arriving at a comparison of the efficacy of the proposed SNR estimation method.

4.5.1 Number of symbols needed for estimation via $\hat{I}_{M,N}$

From (92) or (111) it is clear that in order to achieve a more accurate value of $\hat{I}_{M,N}$, and by extension of $\gamma = f_M^{-1}(\hat{I}_{M,N})$, N should be increased until $\mathrm{var}\left(\hat{I}_{M,N}\right)$ falls below an acceptable value. Indeed, in concordance with the discussion in Sec. 4.4 and (115), the purpose of this section may be stated as follows: we would like to compute the minimal value of $2 \cdot N$ needed to achieve a desired tolerance in the estimation of the E_S/N_0, with a desired confidence. Mathematically, the question is: *What is the minimal value of $2 \cdot N$ needed so that the following holds?*

$$P\left(\left| f_M^{-1}\left(\hat{I}_{M,N}\right) - \chi \right| < tol\right) > C \tag{116}$$

where in (116) *tol* is the tolerance and C is the confidence. Note that we are interested in $2 \cdot N$ (not simply of N) because $2 \cdot N$ is the number of symbols used in computation of the lock metric from which the E_S/N_0 is estimated (see (105)).

Assume *tol* is in units of dB. We define the constants:

$$r_1 = \left(10^{tol/10} - 1\right) \tag{117}$$

and

$$r_2 = \left(1 - 10^{-tol/10}\right) \tag{118}$$

These constants describe the allowed deviations from the E_S/N_0. Since $f_M(\chi)$ is a monotonously increasing function, (116) implies:

$$P\left(-r_2 \cdot \chi < f_M^{-1}\left(\hat{I}_{M,N}\right) - \chi < r_1 \cdot \chi\right) > C$$
$$\Leftrightarrow P\left(f_M\left((1-r_2) \cdot \chi\right) < \hat{I}_{M,N} < f_M\left((1+r_1) \cdot \chi\right)\right) > C. \tag{119}$$

Now, defining:

$$y \triangleq \min\left\{\left| f_M\left((1-r_2) \cdot \chi\right) - f_M(\chi) \right|, \left| f_M\left((1+r_1) \cdot \chi\right) - f_M(\chi) \right|\right\} \tag{120}$$

and using $E\left[\hat{I}_{M,N} \mid E_S/N_0 = \chi\right] = f_M(\chi)$:

$$P\left(f_M\left((1-r_2)\cdot\chi\right)<\hat{i}_{M,N}<f_M\left((1+r_1)\cdot\chi\right)\right)$$

$$\geq P\left(\left|\frac{\hat{i}_{M,N}-E\left[\hat{i}_{M,N}\middle|E_S/N_0=\chi\right]}{\sqrt{\operatorname{var}\left(\hat{i}_{M,N}\right)}}\right|<\frac{y}{\sqrt{\operatorname{var}\left(\hat{i}_{M,N}\right)}}\right) \tag{121}$$

Because $\hat{i}_{M,N}$ is Gaussian, it follows from (121) that in order for (116) to occur it is sufficient to require (see [13 Chap. 2]):

$$P\left(\left|\frac{\hat{i}_{M,N}-E\left[\hat{i}_{M,N}\middle|E_S/N_0=\chi\right]}{\sqrt{\operatorname{var}\left(\hat{i}_{M,N}\right)}}\right|<\frac{y}{\sqrt{\operatorname{var}\left(\hat{i}_{M,N}\right)}}\right)$$

$$=\operatorname{erf}\left(\frac{y}{\sqrt{2}\sqrt{\operatorname{var}\left(\hat{i}_{M,N}\right)}}\right)>C \tag{122}$$

with $\operatorname{erf}(x)=\frac{2}{\sqrt{\pi}}\int_0^x e^{-t^2}dt$. If carrier phase jitter is negligible, then using (111) in (122) and solving for N:

$$2N>2\left(\frac{\operatorname{erf}^{-1}(C)}{y}\right)^2\left(\int_{-\pi}^{\pi}\cos^2(M\Delta\phi)\cdot p_R(\Delta\phi|\chi)\cdot d\Delta\phi-\left(f_M(\chi)\right)^2\right)$$

$$=2\left(\frac{\operatorname{erf}^{-1}(C)}{y}\right)^2\left(\tfrac{1}{2}+\tfrac{1}{2}f_{2M}(\chi)-\left(f_M(\chi)\right)^2\right) \tag{123}$$

and further closed-form simplifications of (123) are possible using (31)-(32). Conversely, a worst-case result accounting for phase error jitter may be obtained from (122) and (109):

$$2N>2\left(\frac{\operatorname{erf}^{-1}(C)}{y}\right)^2 \tag{124}$$

(Note that for all the equations in this subsection $f_M(\chi)$ is given in (110). See comments made in Sec. 4.3.1).

To recap, we have shown that choosing N which complies with (123) or (124) ensures the verity of (116). As previously stated, eqs. (123)-(124) are expressed in terms of $2 \cdot N$ (not simply of N) because $2 \cdot N$ is the number of symbols used in computation of the lock metric from which the E_S / N_0 is estimated (see (105)).

4.5.2   Number of symbols needed for SNR estimation via the *SER*

A great proportion of modern communications systems produce SNR estimates by measuring the pre- or post-decoder error rate. For example, this is what is done in countless systems that estimate the SNR from the number of errors detected in preambles or training sequences that are embedded in the data stream. Thus, perhaps the most meaningful and universally applicable yardstick by which to measure the efficacy of SNR estimation via (113) is attained through comparison of (123) or (124) to the number of symbols needed for E_S / N_0 estimation via measurement of the *pre*-decoder Symbol Error Rate (SER). This is because for coded signals, the post-decoder error rate is always smaller (often by orders of magnitude) than the pre-decoder SER, i.e. the number of symbols needed for SNR estimation via the pre-decoder SER can also be viewed as a lower bound for that which is needed for estimation via post-decoder error rates, regardless of the coding scheme used.

From [13 eq. 5.2-56], the uncoded M-PSK symbol error probability is:

$$P_e(M,\chi) = 1 - \int_{-\pi/M}^{\pi/M} p_R(\Delta\phi \mid \chi) \cdot d(\Delta\phi) \triangleq g_M(\chi) \tag{125}$$

where eq. (107) defines $p_R(\Delta\phi \mid \chi)$ (note that (125) ignores the effects of phase jitter on the SER. Eq. (125) can still model such effects by first incorporating them [19 Sec. 4.3] into a reduced effective E_S / N_0, but Section 4.5.3 shows that the advantage of the method in Section 4.3 is so great that any effects of such a minute correction are irrelevant).

It shall be commented that another equivalent formula has been suggested by Craig [107]:

$$P_e(M,\chi) \triangleq \frac{1}{\pi} \int_0^{\pi - \pi/M} \exp\left(-\chi \cdot \frac{\sin^2(\pi/M)}{\sin^2(\phi)} \right) d\phi \tag{126}$$

The formula of (126) is quite useful and much easier to work with than (125). This is due to the fact that, unlike (125), in (126) the limits in the integral are finite and the integrand is composed entirely of elementary functions. This allow numerical calculations to occur significantly faster.

To avoid cumbersome notation, we denote the left-hand side of (125) or (126) simply as P_e. We define the binary auxiliary variables U_i as $U_i = 1 \Leftrightarrow$ an error was detected in symbol i. Assuming the errors in the received symbols occur randomly and independently, we have $P(U_i = 1) = P_e$ and $P(U_i = 0) = 1 - P_e$, and therefore $E[U_i] = P_e$ and $\text{var}(U_i) = P_e(1 - P_e)$. We can define the measured SER as $S(L) = \frac{1}{L}\sum_{i=1}^{L} U_i$, and from the central limit theorem we then have:

$$S(L) \sim N\left(P_e, \frac{P_e(1-P_e)}{L}\right) \tag{127}$$

Since $E[S(L)] = P_e$ we can estimate the E_s/N_0 from $S(L)$ via:

$$\eta = g_M^{-1}(S(L)). \tag{128}$$

For comparison to (123)-(124), we are interested in finding a value of L that ensures that:

$$P\left(-r_2 \cdot \chi < \eta - \chi < r_1 \cdot \chi\right) > C \tag{129}$$

where $r_1 = \left(10^{tol/10} - 1\right)$ and $r_2 = \left(1 - 10^{-tol/10}\right)$. Developing equation (129) further, we have that it is equivalent to:

$$P\left((1-r_2) \cdot \chi < g_M^{-1}(S(L)) < (1 + r_1) \cdot \chi\right) > C. \tag{130}$$

Note that $g_M(\chi)$ is a monotonically decreasing function [13 Sec. 5.2.7] (in words: the probability of symbol error is inversely related to the E_s/N_0). This is shown in Fig. 68.

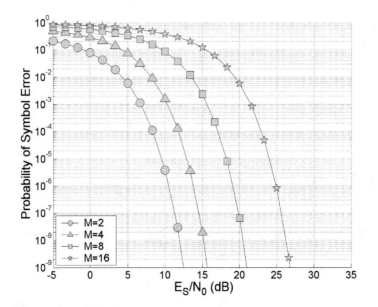

Fig. 68. Probability of symbol error $P_e = g_M(E_s/N_0)$ **as a function of the** E_s/N_0 **ratio.**

Using the monotonically decreasing nature of $g_M(E_s/N_0)$, we have that an equivalent requirement to (130) is:

$$P\big(g_M((1-r_2)\cdot\chi) > S(L) > g_M((1+r_1)\cdot\chi)\big) > C. \tag{131}$$

Defining:

$$z \triangleq \min\big\{\big|g_M((1-r_2)\cdot\chi) - P_e\big|, \big|g_M((1+r_1)\cdot\chi) - P_e\big|\big\} \tag{132}$$

we have that for (131) to be fulfilled it suffices that:

$$P\left(\left|\frac{S(L)-P_e}{\sqrt{\mathrm{var}(S(L))}}\right| < \frac{z}{\sqrt{\mathrm{var}(S(L))}}\right) > C \tag{133}$$

Eq. (127) applied to (133) means that for (129) to be guaranteed it is sufficient to require:

$$L > 2P_e \left(1 - P_e\right) \bigg/ \left(\frac{z}{erf^{-1}(C)}\right)^2 \qquad (134)$$

4.5.3 Display and Analysis of Results

A meaningful appraisal of the utility of estimating the E_s/N_0 via (113) may be obtained by comparing the number of symbols necessary for such an estimate, as per (123) or (124), to that which is required to attain a similarly accurate estimate from the SER, as per (134). We now embark upon making such comparisons.

In Fig. 69, Fig. 70, and Fig. 71 we see a comparison of estimation via $\hat{l}_{M,N}$ vs. estimation via the SER, for BPSK, QPSK, and 8-PSK, for $C = 99\%$ and $C = 99.9\%$, with $tol = 0.5\,\mathrm{dB}$. As those figures clearly illustrates, estimation via $\hat{l}_{M,N}$ is particularly advantageous for higher E_s/N_0 ratios, where it is seen that, while estimation via the SER experiences exponential growth in the number of symbols necessary, the growth rate for estimation from $\hat{l}_{M,N}$ is much milder.

It is worthwhile noting that the fact that the required number of symbols for both estimation methods increases at high SNRs is a manifestation of the fact that in both cases the number of required symbols is asymptotically inversely dependent upon the squared magnitude of the derivative $f_M(\chi)$ and $g_M(\chi)$ (respectively). The fact that the growth rate for estimation from $\hat{l}_{M,N}$ is much milder is a manifestation of the fact that as the SNR increases $\left|\dfrac{d(f_M)}{d\chi}\right|$ tends to 0 at a much milder rate than $\left|\dfrac{d(g_M)}{d\chi}\right|$. This issue is explored in depth in Secs. 4.5.4 and 4.5.5.

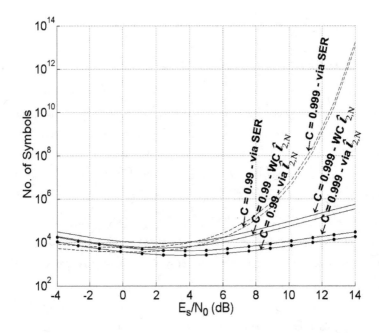

Fig. 69. Number of symbols needed to estimate the E_S/N_0 to within $\pm 0.5\,\mathrm{dB}$, for $M = 2$ (BPSK). "WC" denotes worst-case results for $\hat{I}_{M,N}$, i.e. using (124). Other curves obtained using (123) and (134).

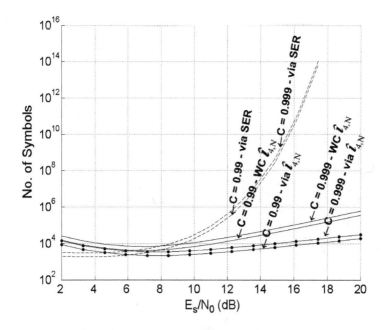

Fig. 70. Number of symbols needed to estimate the E_S/N_0 to within $\pm 0.5\,\mathrm{dB}$, for $M = 4$ (QPSK). "WC" denotes worst-case results for $\hat{I}_{M,N}$, i.e. using (124). Other curves obtained using (123) and (134).

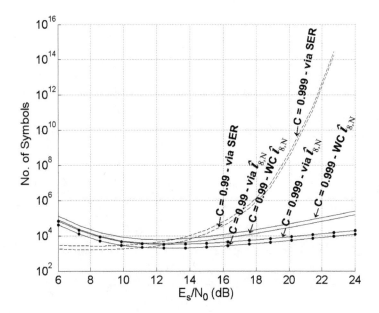

Fig. 71. Number of symbols needed to estimate the E_S/N_0 to within $\pm 0.5\,\text{dB}$, for $M=8$ (8-PSK). "WC" denotes worst-case results for $\hat{I}_{M,N}$, i.e. using (124). Other curves obtained using (123) and (134).

Some comments are in order regarding Fig. 69 to Fig. 71 when the data stream sent is not a known sequence but is rather an arbitrary data stream. In those figures, it is seen that for low E_S/N_0 ratios, apparently estimation via the SER requires less symbols. However this is a fallacy since Fig. 69 to Fig. 71 assume that the SER is measured precisely. The only way to achieve a data-independent SER measurement is by first using a error correction decoder and then comparing the corrected data to the input data stream [62], thus arriving at the pre-decoder SER. This tacitly assumes that the post-decoder data stream is error-free. Yet, at low E_S/N_0 the post-decoder data stream cannot be approximated as error-free, thus inherently skewing the SER measurement. Indeed, the decoder may not even be in lock for low E_S/N_0 ratios, making the SER measurement impossible in the first place. "Training sequences", or known sequences of symbols, can be transmitted in order to arrive at an accurate SER estimate; however this means that at least some of the channel throughput is taken up by such sequences. In contrast, estimation via $\hat{I}_{M,N}$ suffers no such impediments, as it is independent of the data stream, the decoding scheme, and the post-decoder error rate. Thus, Fig. 69 to Fig. 71 can be viewed as optimistic w.r.t. estimation via the SER, and consequently estimation via $\hat{I}_{M,N}$ can be considered superior for all E_S / N_0 ratios. Furthermore, estimation via the SER requires error detection and accrual mechanisms, which often necessitate non-trivial hardware and/or software resource appropriations. This is quite different from the simple and compact hardware implementation of the proposed SNR estimation method, as given in Sec. 4.3.2.

4.5.4   Asymptotic Formulas for Number of Necessary Symbol Intervals

In this section we shall endeavour to find asymptotic closed-form formulas for the number of symbol intervals necessary for estimation of the SNR (i.e., asymptotic formulas for (124) and (134)). As we shall see, these approximations will enable us to attain a better intuitive understanding of the estimators' performance as it was displayed in Section 4.5.3.

Starting off with (124), we note that from (120) that

$$y \triangleq \min\left\{ \left| f_M\left((1-r_2)\cdot\chi\right) - f_M(\chi) \right|, \left| f_M\left((1+r_1)\cdot\chi\right) - f_M(\chi) \right| \right\}$$ and from (117)

and (118) that $r_1 = \left(10^{tol/10} - 1\right)$ and $r_2 = \left(1 - 10^{-tol/10}\right)$. Let us investigate how y behaves for very small values of *tol*.

The Taylor expansion of the expression 10^x is [54 eq. 20.16]:

$$10^x = 1 + x\cdot\ln(10) + \frac{\left(x\cdot\ln(10)\right)^2}{2!} + \dots$$ (135)

Now, if *tol* is very small, we have that we can approximate $10^{tol/10}$ by retaining only the first two terms of the Taylor series, upon which we have that:

$$10^{tol/10} \approx 1 + \frac{tol}{10} \cdot \ln(10) \tag{136}$$

and similarly:

$$10^{-tol/10} \approx 1 - \frac{tol}{10} \cdot \ln(10) \tag{137}$$

This means that from (117) and (118) we have:

$$r_1 = \left(10^{tol/10} - 1\right) \approx 1 + \frac{tol}{10} \cdot \ln(10) - 1 = \frac{tol}{10} \cdot \ln(10) \tag{138}$$

and

$$r_2 = \left(1 - 10^{-tol/10}\right) \approx 1 - \left(1 - \frac{tol}{10} \cdot \ln(10)\right) = \frac{tol}{10} \cdot \ln(10) \tag{139}$$

Thus, we find that for small values of tol we have $r_1 \approx r_2$. To give an example of the validity of this approximation, take $tol = 0.5 \text{ dB}$, which was the case analyzed in Section 4.5.3. For $tol = 0.5 \text{ dB}$ we have $r_1 = \left(10^{tol/10} - 1\right) = 0.122$ and $r_2 = \left(1 - 10^{-tol/10}\right) = 0.108$, and we have that $r_2 / r_1 \times 100 = 89.1\%$. Hence, $r_1 \approx r_2$ is quite a good approximation. For simplicity of notation, we define the variable $r \triangleq \frac{r_1 + r_2}{2}$, then for small tol we have $r_1 \approx r_2 \approx r$. With this notation we see from (120) that:

$$
\begin{aligned}
y &\triangleq \min\left\{ \left| f_M\left((1-r_2) \cdot \chi\right) - f_M(\chi) \right|, \left| f_M\left((1+r_1) \cdot \chi\right) - f_M(\chi) \right| \right\} \\
&\approx \min\left\{ \left| f_M\left((1-r) \cdot \chi\right) - f_M(\chi) \right|, \left| f_M\left((1+r) \cdot \chi\right) - f_M(\chi) \right| \right\} \\
&= r\chi \cdot \min\left\{ \left| \frac{f_M\left((1-r) \cdot \chi\right) - f_M(\chi)}{r\chi} \right|, \left| \frac{f_M\left((1+r) \cdot \chi\right) - f_M(\chi)}{r\chi} \right| \right\} \tag{140} \\
&\approx r\chi \cdot \left| \frac{d(f_M)}{d\chi} \right|
\end{aligned}
$$

where we have used the fact that $r > 0$, that r is very small, and that f_M is continuous so that:

$$\lim_{r \to 0} \left| \frac{f_M\left((1-r)\cdot\chi\right) - f_M\left(\chi\right)}{r\chi} \right| = \lim_{r \to 0} \left| \frac{f_M\left((1+r)\cdot\chi\right) - f_M\left(\chi\right)}{r\chi} \right| = \left| \frac{d\left(f_M\right)}{d\chi} \right| \tag{141}$$

Continuing, we find from (140) and (124) that:

$$2N > 2\left(\frac{erf^{-1}(C)}{y} \right)^2 \approx 2\left(\frac{erf^{-1}(C)}{r\chi} \right)^2 \Big/ \left(\left| \frac{d\left(f_M\right)}{d\chi} \right| \right)^2 \tag{142}$$

Now, using $f_M\left(\chi\right) \approx \exp\left(\frac{-M^2}{4\chi}\right)$ (see (35)) we find that

$$\left| \frac{d\left(f_M\right)}{d\chi} \right| \approx \frac{M^2}{4}\exp\left(\frac{-M^2}{4\chi}\right) \cdot \left| \frac{d\left(1/\chi\right)}{d\chi} \right| = \frac{M^2}{4\chi^2}\exp\left(\frac{-M^2}{4\chi}\right) \tag{143}$$

Plugging (143) into (142) we find that:

$$2N > 2\left(\frac{erf^{-1}(C)}{y} \right)^2 \approx 2\left(\frac{erf^{-1}(C)}{r\chi} \right)^2 \Big/ \left(\frac{M^2}{4\chi^2}\exp\left(\frac{-M^2}{4\chi}\right) \right)^2 \tag{144}$$

At high SNR, i.e. for $\chi \to \infty$, we have from (144) that:

$$2N > 2\left(\frac{erf^{-1}(C)}{y} \right)^2 \approx 2\left(\frac{erf^{-1}(C)}{r\chi} \right)^2 \Big/ \left(\frac{M^2}{4\chi^2}\exp\left(\frac{-M^2}{4\chi}\right) \right)^2$$
$$\underset{\chi \to \infty}{\approx} 2\left(\frac{erf^{-1}(C)}{r} \right)^2 \cdot \frac{16\chi^2}{M^4} \tag{145}$$

Now, let us perform the same procedure for estimation via the SER and $g_M\left(\chi\right)$. We find from (132) that:

$$z \triangleq \min\left\{ \left| g_M\left((1-r_2)\cdot\chi\right) - g_M\left(\chi\right) \right|, \left| g_M\left((1+r_1)\cdot E_S/N_0\right) - g_M\left(\chi\right) \right| \right\}$$
$$\approx r\chi\left| \frac{d(g_M)}{d\chi} \right| \tag{146}$$

(where we used the fact that $g_M\left(\chi\right) = P_e$ (see (125))) and thus from (134) we find that:

$$L > 2P_e\left(1 - P_e\right) \Bigg/ \left(\frac{z}{erf^{-1}(C)}\right)^2$$

$$\approx 2g_M\left(\chi\right)\left(1 - g_M\left(\chi\right)\right) \Bigg/ \left(\frac{r\chi\left|\dfrac{d(g_M)}{d\chi}\right|}{erf^{-1}(C)}\right)^2 \qquad (147)$$

$$= 2\left(\frac{erf^{-1}(C)}{r\chi}\right)^2 \cdot \frac{\left(g_M\left(\chi\right) - g_M^{\ 2}\left(\chi\right)\right)}{\left|\dfrac{d(g_M)}{d\chi}\right|^2}$$

From [13 eq. 5.2-61] we have that $g_M(\chi) \approx 2 \cdot Q\left(\sqrt{2\chi}\sin\dfrac{\pi}{M}\right)$ where [13 eqs. 2.1-97,

2.1-98] $Q(x) \triangleq \dfrac{1}{\sqrt{2\pi}}\int_x^\infty e^{-t^2/2}dt = \dfrac{1}{2}erfc\left(\dfrac{x}{\sqrt{2}}\right)$. Furthermore, from [19 App. 3B] we have

at high SNR

$$Q(x) \overset{x\to\infty}{\approx} \frac{1}{x\sqrt{2\pi}}\exp\left(\frac{-x^2}{2}\right) \qquad (148)$$

so that:

$$g_M\left(\chi\right) \approx 2 \cdot Q\left(\sqrt{2\chi}\sin\frac{\pi}{M}\right)$$

$$\approx 2 \cdot \left[\frac{1}{x\sqrt{2\pi}}\exp\left(\frac{-x^2}{2}\right)\right]_{x=\sqrt{2\chi}\sin\frac{\pi}{M}} \qquad (149)$$

$$\approx \frac{1}{\sqrt{\chi \cdot \pi} \cdot \sin\dfrac{\pi}{M}}\exp\left(-\chi \cdot \sin^2\frac{\pi}{M}\right)$$

and, moreover:

$$\frac{dQ(x)}{dx} \overset{x \to \infty}{\approx} \frac{d}{dx}\left(\frac{1}{x\sqrt{2\pi}} \exp\left(\frac{-x^2}{2} \right) \right)$$

$$= \frac{-1}{x^2\sqrt{2\pi}} \exp\left(\frac{-x^2}{2} \right) + \frac{-x}{x\sqrt{2\pi}} \exp\left(\frac{-x^2}{2} \right) \qquad\qquad (150)$$

$$\overset{x \to \infty}{\approx} \frac{-1}{\sqrt{2\pi}} \exp\left(\frac{-x^2}{2} \right)$$

and therefore:

$$\frac{d\left(g_M(\chi) \right)}{d\chi} \approx 2 \cdot \frac{d}{d\chi} Q\left(\sqrt{2\chi} \sin\frac{\pi}{M} \right)$$

$$= 2 \cdot \left[\frac{d}{dx} Q(x) \right]_{x=\sqrt{2\chi}\sin\frac{\pi}{M}} \cdot \frac{d}{d\chi}\left(\sqrt{2\chi} \sin\frac{\pi}{M} \right)$$

$$= 2 \cdot \left[\frac{d}{dx} Q(x) \right]_{x=\sqrt{2\chi}\sin\frac{\pi}{M}} \cdot \frac{1}{\sqrt{2\chi}} \sin\frac{\pi}{M} \qquad\qquad (151)$$

$$\approx 2 \cdot \frac{-1}{\sqrt{2\pi}} \exp\left(-\chi \sin^2 \frac{\pi}{M} \right) \cdot \frac{1}{\sqrt{2\chi}} \sin\frac{\pi}{M}$$

$$= \frac{-1}{\sqrt{\chi \cdot \pi}} \exp\left(-\chi \sin^2 \frac{\pi}{M} \right) \cdot \sin\frac{\pi}{M}$$

Thus, using (149) and (151):

$$\frac{\left(g_M(\chi) - g_M{}^2(\chi)\right)}{\left|\dfrac{d(g_M)}{d\chi}\right|^2} \underset{\chi \to \infty}{\approx} \frac{g_M(\chi)}{\left|\dfrac{d(g_M)}{d\chi}\right|^2}$$

$$\underset{\chi \to \infty}{\approx} \frac{\dfrac{1}{\sqrt{\chi \cdot \pi} \cdot \sin \dfrac{\pi}{M}} \exp\left(-\chi \cdot \sin^2 \dfrac{\pi}{M}\right)}{\left(\dfrac{1}{\sqrt{\chi \cdot \pi}} \exp\left(-\chi \sin^2 \dfrac{\pi}{M}\right) \cdot \sin \dfrac{\pi}{M}\right)^2} \tag{152}$$

$$= \frac{\sqrt{\chi \cdot \pi} \cdot \exp\left(\chi \cdot \sin^2 \dfrac{\pi}{M}\right)}{\sin^3 \dfrac{\pi}{M}}$$

and therefore from (147):

$$L > 2 P_e \left(1 - P_e\right) \Big/ \left(\frac{z}{erf^{-1}(C)}\right)^2$$

$$\approx 2 \left(\frac{erf^{-1}(C)}{r\chi}\right)^2 \cdot \frac{\left(g_M(\chi) - g_M{}^2(\chi)\right)}{\left|\dfrac{d(g_M)}{d\chi}\right|^2}$$

$$\approx 2 \left(\frac{erf^{-1}(C)}{r\chi}\right)^2 \cdot \frac{\sqrt{\chi \cdot \pi} \cdot \exp\left(\chi \cdot \sin^2 \dfrac{\pi}{M}\right)}{\sin^3 \dfrac{\pi}{M}} \tag{153}$$

$$= 2 \left(\frac{erf^{-1}(C)}{r}\right)^2 \cdot \frac{\chi^{-3/2} \sqrt{\pi} \cdot \exp\left(\chi \cdot \sin^2 \dfrac{\pi}{M}\right)}{\sin^3 \dfrac{\pi}{M}}$$

Let us take a look at (145) and (153). We see that the dependence of $2N$ upon the SNR χ is *polynomial*, while the dominant dependence of L upon χ is *exponential*. Since an exponential dependence will always grow much more rapidly than any(finite order polynomial dependence, then we have just proven a theoretical justification for the results presented in Fig. 69 to Fig. 71.

To verify the derivations of this section, we shall compute and graph (145) and (153) vs. the results obtained from (124) and (134). This is done in Fig. 72 to Fig. 74, for $tol = 0.5$ dB and $C = 99\%$. As can be seen from those figures, the asymptotic expressions are quite accurate and hence provide a useful tool that the designer can use to roughly calculate the estimation interval requirements at high SNR. Fig. 72 to Fig. 74 also serve to validate the theoretical calculations made in this section, the results presented in Section 4.5.3, and the computer algorithms used to generate those results.

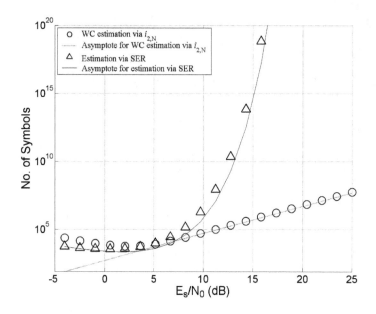

Fig. 72. Number of symbols needed to estimate the E_s/N_0 to within ± 0.5 dB, with a confidence of C=99%, for $M = 2$ (BPSK). "WC" denotes worst-case results for $\hat{i}_{M,N}$, i.e. using (124). Results for estimation vs. the SER obtained via (134). Asymptotic results obtained from (145) and (153).

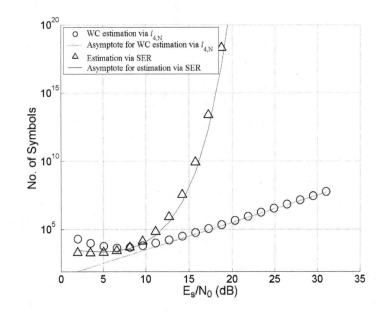

Fig. 73. Number of symbols needed to estimate the E_S/N_0 to within $\pm 0.5\,\mathrm{dB}$, with a confidence of C=99%, for $M=4$ (QPSK). "WC" denotes worst-case results for $\hat{l}_{M,N}$, i.e. using (124). Results for estimation vs. the SER obtained via (134). Asymptotic results obtained from (145) and (153).

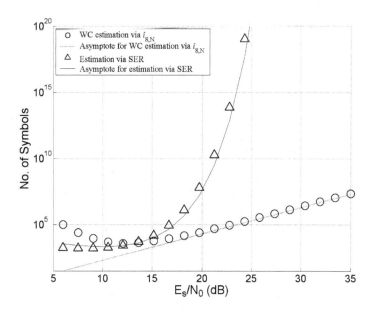

Fig. 74. Number of symbols needed to estimate the E_S/N_0 **to within** $\pm 0.5\,\text{dB}$**, with a confidence of C=99%, for** $M=8$ **(8-PSK). "WC" denotes worst-case results for** $\hat{\imath}_{M,N}$**, i.e. using (124). Results for estimation vs. the SER obtained via (134). Asymptotic results obtained from (145) and (153).**

4.5.5   The Causes of the Proposed Method's Advantages

Let us now recapacitate and investigate further the implementation-related and performance-related differences between estimation via the SER and estimation via $\hat{\imath}_{M,N}$, with the aim of better understanding the roots of the advantage of the proposed method. In practical receivers, neither f_M^{-1} nor g_M^{-1} would be computed directly; rather, they would be computed beforehand and stored in a lookup table. This is true even for software implementations. This is due to the fact that there is no closed-form formula for the inverses of either function, and so this must be done numerically. Doing such numerical calculations for each estimate would simply take too long and would preclude any real time operation. Thus, the results would be computed beforehand, and stored either in a lookup table (in a hardware implementation) or in an array stored in the computer's memory (in a software implementation). Regarding the fundamental need to compute both inverse functions in this manner, there is no essential difference between estimation via the SER and estimation via $\hat{l}_{M,N}$. The real differences between the methods are that:

1. The lookup table (or memory array) for computation of f_M^{-1} would be much smaller than that for g_M^{-1}, at least in a fixed-point implementation (which basically covers all hardware implementations in an ASIC or FPGA)

2. Estimation via $\hat{l}_{M,N}$ does not include the need to perform error detection, which in contrast is a prerequisite for estimation via g_M^{-1}.

3. As the analysis in the chapter shows, estimation via $\hat{l}_{M,N}$ requires significantly less symbols to arrive an equally accurate estimate.

Let us analyze the above points individually.

1. The lookup table for f_M^{-1} is much smaller because its input, which is $\hat{l}_{M,N}$, always lies in the interval $[0,1]$ and is *"well-behaved"* as a function of the χ_{dB} (remember, estimation is only possible when the carrier loop is locked, whereupon the output of the lock detector is in the interval $[0,1]$). The meaning of "well-behaved" is that the derivative of $f_M(\chi) = E\left[\hat{l}_{M,N} \mid E_S / N_0 = \chi\right]$ as a function of χ_{dB} has a small absolute value, i.e. $\left|\dfrac{d\left(f_M(\chi)\right)}{d\left(\chi_{dB}\right)}\right|$ is small for "reasonable" χ_{dB} (i.e., SNRs above the threshold of the respective modulation,

- 158 -

but not excessively high). A manifestation of this phenomenon is that $\left| \dfrac{d\left(f_M^{-1} \right)}{d\left(\hat{l}_{M,N} \right)} \right|$

is also "well-behaved", and this can be easily seen from inspection of Fig. 66. Since $\hat{l}_{M,N}$ behaves "nicely" as a function of χ_{dB}, the input of the lookup table computing f_M^{-1} in a fixed-point implementation does not need to be quantized to any great precision in order to achieve accurate estimation of all practical χ_{dB}. This is in sharp contrast to a lookup table used for computation of g_M^{-1}. The input of that lookup table is the SER, which, while also in the interval $[0,1]$, is by contrast is not as "well behaved" because it tends towards zero very fast for moderate and high χ_{dB} (see Fig. 68). This means that if any accuracy is desired for estimation of those χ_{dB} ratios via g_M^{-1}, then the input of the lookup table implementing g_M^{-1} needs to be able to accurately represent very small numbers (for moderate and high χ) which differ in orders of magnitude from each other as the χ_{dB} increases only very slightly. Thus, in a fixed-point implementation this means that many more bits are needed for the input of the lookup table that implements g_M^{-1}. Now, denote the number of bits at the input of the lookup table as k, then the number of entries in the lookup table is 2^k. So, the size of the lookup table implementing g_M^{-1} in a fixed-point architecture will be much larger than that needed for implementation of f_M^{-1}.

2. This point is rather obvious and was discussed in Sec. 4.5.3. Of course, the implementation of error correction hardware or software is no trivial matter, and the lack of it for the proposed method is an outright saving of resources. Also, as noted in Sec. 4.5.3, at low SNRs the error correction decoder may operate with a non-negligible output error rate (if it is locked at all), thus hampering estimation via g_M^{-1}.

3. The fact that the proposed estimation method requires much less symbols than that required for estimation via g_M^{-1} is clearly seen from the results developed and

presented in Sec. 4.5.3. This means that estimation via f_M^{-1} is much better suited for real-time estimate generation.

4.6 Comparison of Estimation via $\hat{l}_{M,N}$ to additional SNR estimation methods

4.6.1 Qualitative comparison

Several additional SNR estimation methods are presented in [59], [60], [61], [62], [63], [64], [65], [66], [67], [68], [69], [70], [71], [72], [73], [16] and [74]. While a quantitative and exhaustive comparison versus all of those methods could be undertaken, this is unwarranted since the qualitative characteristics of the proposed SNR estimator make it so attractive that it renders such a comparison unnecessary. To this end, it shall be commented that of the [59], [60], [61], [62], [63], [64], [65], [66], [67], [68], [69], [70], [71], [72], [73], [16] and [74]:

- Some are unique to a specific receiver structure
- Most require some form of symbol decisions to be made
- Most require more than one sample per symbol
- (Most importantly:) **None** of those methods appear to have a hardware implementation **nearly** as compact as the one suggested here.

Moreover, it is mentioned that, due to the resilience of $\hat{l}_{M,N}$ to effects stemming from imperfect AGC control of K in Fig. 13 (see Chapter 2), this resilience is also present in the resulting E_s/N_0 estimates, thus deeming them reliable even when rapidly fading signal conditions are encountered (see also discussion in Sec. 4.7).

4.6.2 Quantitative comparison of NMSE vs. other blind 1-sample/symbol SNR estimators

Although, as stated above, it would be impossible to engage in quantitative comparisons with all of the SNR estimators referenced in Sec. 4.6.1, we shall, for completeness, engage in quantitative comparisons vs. the M_2M_4 estimator and the Signal-to-Variation Ratio (SVR) estimator which are analyzed in [60]. This is because the performance of the aforementioned estimators (especially the M₂M₄ estimator) has been found in [60] to be very good (in fact, the M₂M₄ estimator is judged one of the "best" SNR estimators [60 Sec. V]). Moreover, the M₂M₄ and SVR estimators are blind methods that need a sampling rate of 1 sample/symbol, which are exactly the characteristics of the proposed SNR estimator. Hence, a comparison is appropriate.

First, following [60], let us review the M₂M₄ and SVR estimators.

a)   The M₂M₄ estimator

Define $r_n \triangleq I(n) + j \cdot Q(n)$. The M₂M₄ estimator utilizes the 2nd-moment and 4th-moment of the signal, defined as:

$$M_2 \triangleq E\left[\left|r_n\right|^2\right] \tag{154}$$

and:

$$M_4 \triangleq E\left[\left|r_n\right|^4\right] \tag{155}$$

It can be shown [60 eq. (35)-(36)] that we have:

$$M_2 = S + N \tag{156}$$

and:

$$M_4 = k_a S^2 + 4SN + k_w N^2 \tag{157}$$

where S is the signal power, N is the noise variance, $k_a \triangleq E\left[\left|a_n\right|^4\right] / \left(E\left[\left|a_n\right|^2\right]\right)^2$ is the signal kurtosis and $k_w \triangleq E\left[\left|n_n\right|^4\right] / \left(E\left[\left|n_n\right|^2\right]\right)^2$ is the noise kurtosis (where we define $n_n \triangleq n_I(n) + j \cdot n_Q(n)$). We can solve (156)-(157) for S and N to yield [60 eq. (37)-(38)]:

$$\hat{S} = \frac{M_2(k_w - 2) - \sqrt{(4 - k_a k_w)M_2^2 + M_4(k_a + k_w - 4)}}{k_a + k_w - 4} \tag{158}$$

and:

$$\hat{N} = M_2 - \hat{S} \tag{159}$$

For the M-PSK signals discussed in this book, it can be shown [60] that $k_a = 1$ and $k_w = 2$, and using (158)-(159), we find the M₂M₄ SNR estimate as [60 eq. (39)]:

$$\gamma_{M_2 M_4} \triangleq \hat{S} / \hat{N} = \frac{\sqrt{2M_2^2 - M_4}}{M_2 - \sqrt{2M_2^2 - M_4}} \tag{160}$$

Obviously, in an actual implementation the moments M_2 and M_4 are estimated from a finite number of symbols, as follows (from [60 eq. (43)-(44)], with adaptation to the notations used in this book):

$$M_2 = \frac{1}{2N} \sum_{n=-N+1}^{N} |r_n|^2 \tag{161}$$

and:

$$M_4 = \frac{1}{2N} \sum_{n=-N+1}^{N} |r_n|^4 \tag{162}$$

Eq. (160), with the moments M_2 and M_4 computed as in (161)-(162), is the M₂M₄ estimator.

b) The SVR estimator

To define the SVR estimator, we define the auxiliary variable [60 eq. (45)]

$$\beta \triangleq \frac{E\left[|r_n|^2 |r_{n-1}|^2\right]}{E\left[|r_n|^4\right] - E\left[|r_n|^2 |r_{n-1}|^2\right]} \tag{163}$$

The SVR estimator is defined as [60 eq. (48)]:

$$\gamma_{SVR} = \frac{(\beta - 1) + \sqrt{(\beta - 1)^2 - [1 - \beta(k_a - 1)][1 - \beta(k_w - 1)]}}{1 - \beta(k_a - 1)} \tag{164}$$

For the M-PSK signals discusses in this book, it can be shown [60] that $k_a = 1$ and $k_w = 2$, so that (164) simplifies to:

$$\gamma_{SVR} = \beta - 1 + \sqrt{\beta(\beta - 1)} \tag{165}$$

In practice, β is estimated from a finite number of symbols, as follows (from [60 eq. (53)], with adaptation to the notations used in this book):

$$\beta = \frac{\dfrac{1}{2N-1} \sum\limits_{n=-N+2}^{N} |r_n|^2 |r_{n-1}|^2}{\dfrac{1}{2N-1} \sum\limits_{n=-N+2}^{N} |r_n|^4 - \dfrac{1}{2N-1} \sum\limits_{n=-N+2}^{N} |r_n|^2 |r_{n-1}|^2} \qquad (166)$$

Eq. (165), with β computed as in (166), is the SVR estimator.

c)   *Comparison metrics vs. the M_2M_4 and SVR estimators*

In order to facilitate the comparison between the proposed estimator and the M_2M_4 and SVR estimators, we shall use the same metrics used in [60].

The first of these metrics is [60 eq. (65)] Normalized Mean Squared Error (NMSE), i.e. we shall compare $E\left[(\gamma - \chi)^2\right]/\chi^2$, which is the NMSE for the proposed method (with $\gamma \triangleq f_M^{-1}(\hat{l}_{M,N})$, see (112)), with the NMSEs for the M_2M_4 and SVR estimators, which are $E\left[(\gamma_{M_2M_4} - \chi)^2\right]/\chi^2$ and $E\left[(\gamma_{SVR} - \chi)^2\right]/\chi^2$, respectively.

The second metric is the normalized bias [60 eq. (68)], i.e. we shall compare $E\left[(\gamma - \chi)\right]/\chi$ to $E\left[(\gamma_{M_2M_4} - \chi)\right]/\chi$ and $E\left[(\gamma_{SVR} - \chi)\right]/\chi$.

In order to help us evaluate the results, we shall also look at the Cramér-Rao (CRB) bound for the NMSE, which is [60 eq. (65)] :

$$CRB = \frac{2}{\chi \cdot 2N} + \frac{1}{2N} \qquad (167)$$

The CRB is the lowest theoretically attainable NMSE. As for the best theoretically attainable bias, obviously that limit is 0.

In order to understand the significance of the NMSE metric, it is helpful to note that the MSE is $MSE = \chi^2 \cdot NMSE$. Thus, if we have, for example, an NMSE of 10^{-2}, this would mean that the average RMS error of the estimation would be $\sqrt{MSE} = \sqrt{\chi^2 \cdot 10^{-2}} = 0.1\chi$, which in dB is a positive RMS deviation of $10 \cdot \log_{10}(1.1\chi / \chi) = 0.41$ dB and a negative RMS deviation of

$10 \cdot \log_{10}(0.9\chi/\chi) = -0.46$ dB, which is a respectably small error. Even an NMSE of $5 \cdot 10^{-2}$, which will have an RMS error of 0.22χ, which is a positive RMS deviation of $10 \cdot \log_{10}(1.22\chi/\chi) = 0.86$ dB and a negative RMS deviation of $10 \cdot \log_{10}(0.78\chi/\chi) = -1.08$ dB, which is still satisfactory for many receivers.

d) Quantitative results and discussion

In the following figures we shall present NMSE and normalized bias results calculated through simulations. The figures are followed with a discussion of the results.

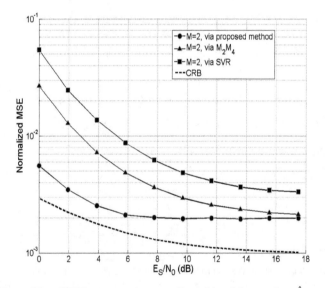

Fig. 75. NMSE comparison of estimation via $\hat{l}_{M,N}$, the M₂M₄ estimator, and the SVR estimator, with $2N = 1024$ symbols used to compute each estimator. Modulation is BPSK ($M = 2$).

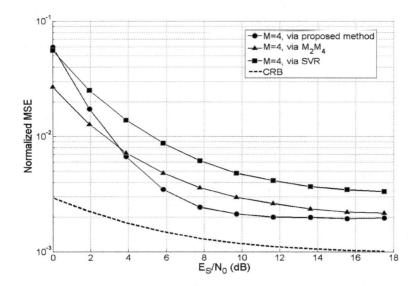

Fig. 76. NMSE comparison of estimation via $\hat{l}_{M,N}$, the M₂M₄ estimator, and the SVR estimator, with $2N = 1024$ symbols used to compute each estimator. Modulation is QPSK ($M = 4$).

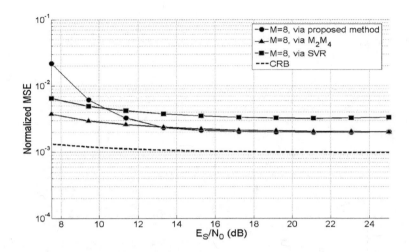

Fig. 77. NMSE comparison of estimation via $\hat{l}_{M,N}$, the M_2M_4 estimator, and the SVR estimator, with $2N = 1024$ symbols used to compute each estimator. Modulation is 8-PSK ($M = 8$).

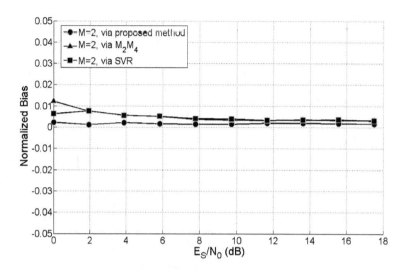

Fig. 78. Normalized bias comparison of estimation via $\hat{l}_{M,N}$, the M_2M_4 estimator, and the SVR estimator, with $2N = 1024$ symbols used to compute each estimator. Modulation is BPSK ($M=2$).

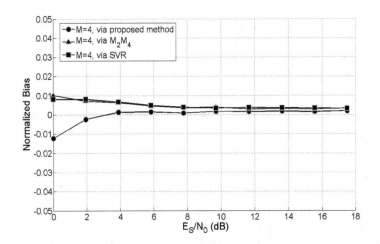

Fig. 79. Normalized bias comparison of estimation via $\hat{l}_{M,N}$, the M_2M_4 estimator, and the SVR estimator, with $2N = 1024$ symbols used to compute each estimator. Modulation is QPSK ($M = 4$).

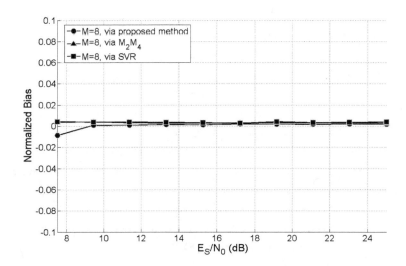

Fig. 80. Normalized bias comparison of estimation via $\hat{l}_{M,N}$, the M₂M₄ estimator, and the SVR estimator, with $2N = 1024$ symbols used to compute each estimator. Modulation is 8-PSK ($M = 8$).

As comparison of the NMSE results in Fig. 75-Fig. 77 shows, the proposed estimator perform better or on par with the M₂M₄ and SVR estimators. The best NMSE results for the proposed method are attained for BPSK, where the proposed method is seen to be better than the M₂M₄ and SVR estimators at all SNRs. For QPSK and 8-PSK the NMSE of the proposed method is best at medium and high SNR, though at the lowest SNRs there is some mild increase in NMSE.

As for bias results, as Fig. 78-Fig. 80 show, all three methods (the proposed method, the M₂M₄ and SVR estimators) have small bias that is very near the optimal value of 0.

To conclude the quantitative comparisons vs. the M₂M₄ and SVR estimators, we have seen that the proposed SNR estimation method is competitive and sometimes superior to those methods.

Some concluding qualitative remarks are in order. First, the proposed method has a very compact hardware implementation that is much more compact than that of the

- 169 -

M_2M_4 and SVR estimators (see [60], [68]), and yet (as we have shown) has comparable and sometimes better performance. Secondly, the M_2M_4 and SVR estimators will not work well in the presence of fading [70], and we shall discuss this in Sec. 4.7. In contrast, the proposed estimator is easily adaptable to work in fading conditions without any appreciable hardware complexity or performance penalties. Finally, it should be noted that the M_2M_4 and SVR estimators, being moment-base estimators, will be able to provide an estimate if the carrier PLL is unlocked, and this will not be true for the proposed method (which requires prior carrier PLL synchronization). However, we shall see in Part B of this chapter that the proposed method can be modified to work in the absence of carrier synchronization and yet retain similar performance.

4.7   SNR Estimation via $\hat{l}_{M,N}$ for M-PSK in the Presence of Fading: Comparison with Estimation via the SER

Until this point in the chapter we have ignored the possible presence of signal fading in the channel. To incorporate such effects into the analysis, we shall assume that our signal is subject to frequency-flat (= frequency-nonselective) slow fading (for discussion of fading processes, see for example [13 Chap. 14]). We shall comment that, in general, a suppressed carrier coherent M-PSK system would perform poorly in frequency-selective or fast fading (i.e. where $T_{COH} \ll T$ or $1/T \ll W_{COH}$) In such a case a different, probably noncoherent, multicarrier, or spread-spectrum type modulation would be chosen (see see [13 p. 818], [78], [79]). Thus, the assumption of frequency-flat slow fading will be appropriate for the overwhelming majority of suppressed-carrier M-PSK systems.

We use the notation χ to refer to the instantaneous SNR, and the notation $\bar{\chi}$ to denote the *average* SNR (i.e., $\bar{\chi} \triangleq E[E_S / N_0]$). The conditional pdf (probability density function)   of   the   SNR   due   to   fading   will   be   denoted   as $p_F\left(\chi | \bar{\chi}\right) \triangleq p\left(E_S / N_0 = \chi | E[E_S / N_0] = \bar{\chi}\right)$. For example, from [77 Table 2] we have for

Nakagami-m fading $p_F\left(\chi | \bar{\chi}\right) = \dfrac{m^m \chi^{m-1}}{\bar{\chi}^m \Gamma(m)} \exp\left(\dfrac{-m\chi}{\bar{\chi}}\right)$.

We differentiate between two cases: (a) $2NT \ll T_{COH}$ and (b) $2NT \gg T_{COH}$. For case (a), during the averaging over $2NT$ symbol intervals that is done in eq. (105) the channel SNR will not have changed much; hence SNR estimation from $\hat{l}_{M,N}$ will yield an estimate of the *instantaneous* SNR ratio χ. For case (b), since $2NT$ is much larger than the channel coherence time, SNR estimation from $\hat{i}_{M,N}$ will yield an estimate of the *average* SNR ratio $\overline{\chi}$. In general, we would ideally like to produce SNR estimates which are instantly available and can be fed in real-time to the decoder, equalizer, or other receiver components which could make good use of them. Hence, ideally, we would be served by perfect knowledge of the instantaneous SNR ratio χ (which, if desired, could be averaged over time in order to produce an estimate of $\overline{\chi}$). However, estimation of χ is not always possible, due to the fact that T_{COH} may be too short as compared to the estimation period $2NT$ which is necessary in order to achieve an acceptable accuracy in the SNR estimation (see discussion in Sec. 2.7 and the following subsections). Nonetheless, if $2NT \gg T_{COH}$, timely knowledge of the average SNR $\overline{\chi}$ is often sufficient in order to facilitate substantial performance gains (e.g. [12], [15]).

4.7.1   Case (a): $2NT \ll T_{COH}$

For the case of $2NT \ll T_{COH}$ the effects of fading upon the SNR are not felt during the estimation period of $2NT$. Thus, we can treat this case as if no fading were present. Hence, the discussion up to now in this part of the chapter has addressed this case, and the results have been given in Secs. 4.4 to 4.6. The SNR estimate generated in this case would be that of the instantaneous SNR χ.

4.7.2   Case (b): $2NT \gg T_{COH}$

For the case of $2NT \gg T_{COH}$, we must take into account the fading distribution. To estimate the SNR ratio $\overline{\chi}$ we first compute the expected value of $\hat{i}_{M,N}$ in the presence of fading, as per (58) which is repeated here:

$$\overline{f}_M(\overline{\chi}) \triangleq E\left[\hat{i}_{M,N}\left| E\left[\frac{E_S}{N_0}\right] = \overline{\chi}\right.\right] =$$
$$\int_0^\infty \left(\int_{-\pi}^\pi \cos(M\Delta\phi)p\left(\Delta\phi\left|\frac{E_S}{N_0} = \chi\right.\right)d(\Delta\phi)\right)p_F(\chi|\overline{\chi})d\chi \tag{168}$$

We can then estimate $\overline{\chi}$ in a way analogous to (113), i.e. via:

$$\overline{\gamma}_{dB} = 10\log_{10}\left(\overline{f}_M^{-1}(\hat{i}_{M,N})\right) \tag{169}$$

- 171 -

In a way analogous to Sec. 4.5 and (124), we can conclude that the amount of symbols needed to estimate $\overline{\chi}$ from $\hat{l}_{M,N}$ is given by (assuming a worst-case scenario jitter-wise):

$$2N > 2 \cdot \frac{\left(erf^{-1}(C)\right)^2}{\left(\min\left\{\left|\overline{f}_M\left((1-r_2)\overline{\chi}\right) - \overline{f}_M(\overline{\chi})\right|, \left|\overline{f}_M\left((1+r_1)\overline{\chi}\right) - \overline{f}_M(\overline{\chi})\right|\right\}\right)^2} \quad (170)$$

As a comparison yardstick by which to measure the efficacy of estimation via (124), we can compare (170) to estimation via the SER. To do this, we first note that the SER for fading channels is (using (126)):

$$\overline{g}_M(\overline{\chi}) \triangleq \int_0^\infty \left(\frac{1}{\pi} \int_0^{\pi - \pi/M} \exp\left(-\chi \cdot \frac{\sin^2(\pi/M)}{\sin^2(\theta)}\right) d\theta\right) p_F(\chi|\overline{\chi}) d\chi \quad (171)$$

Presenting results for all possible fading distributions would be impossible. In the sequel we thus present results for Nakagami-m fading, with the understanding that results for other fading statistics can be obtained in an analogous manner.

For Nakagami-*m* fading we have [108 eqs. (3), (9)] that:

$$\overline{g}_M(\overline{\chi}) \triangleq \frac{1}{\pi} \int_0^{\pi - \pi/M} \left(1 + \frac{\sin^2(\pi/M)}{\sin^2 \xi} \cdot \frac{\overline{\chi}}{m}\right)^{-m} d\xi \quad (172)$$

Graphs $\overline{g}_M(\overline{\chi})$ for Nakagami-*m* fading are given in Fig. 81 to Fig. 84.

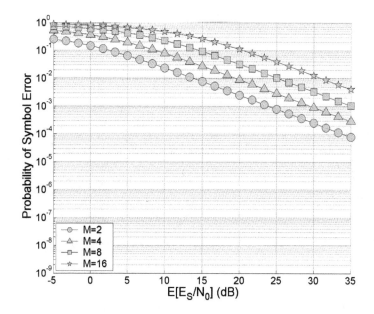

Fig. 81. Probability of symbol error $\overline{g}_M(\overline{\chi})$ as a function of $\overline{\chi}$ for Nakagami-m fading with $m=1$.

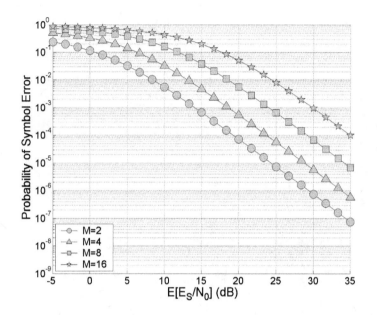

Fig. 82. Probability of symbol error $\overline{g}_M\left(\overline{\chi}\right)$ as a function of $\overline{\chi}$ for Nakagami-m fading with $m=2$.

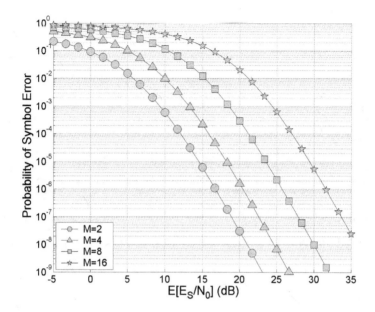

Fig. 83. Probability of symbol error $\overline{g}_M\left(\overline{\chi}\right)$ as a function of $\overline{\chi}$ for Nakagami-m fading with $m = 5$.

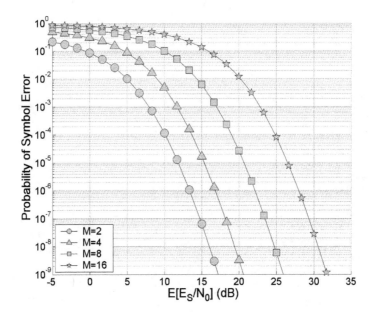

Fig. 84. Probability of symbol error $\overline{g}_M(\overline{\chi})$ as a function of $\overline{\chi}$ for Nakagami-m fading with $m = 10$.

The estimate of the average SNR from the SER, when fading is present, done through:

$$\overline{\eta}_{dB} = 10\log_{10}\left(\overline{g}_M^{-1}(\hat{l}_{M,N})\right) \tag{173}$$

which will give an estimate of $\overline{\chi}$.

Following along the lines of the derivations in Sec. 4.5.2, we find that the number of symbols necessary to estimate $\overline{\chi}$ from the SER (using (173)) is:

$$L > \frac{2\overline{g}_M(\overline{\chi})(1 - \overline{g}_M(\overline{\chi}))(erf^{-1}(C))^2}{\left(\min\left\{\left|\overline{g}_M((1-r_2)\cdot\overline{\chi}) - \overline{g}_M(\overline{\chi})\right|, \left|\overline{g}_M((1+r_1)\cdot\overline{\chi}) - \overline{g}_M(\overline{\chi})\right|\right\}\right)^2} \tag{174}$$

We note that, just as was the case for estimation from $\hat{l}_{M,N}$, we must differentiate between two cases, case (i): $L \ll T_{COH}$ and case (ii): $L \gg T_{COH}$. For case (i) we should use (134) and the estimate will be of χ. For case (ii) we must use (174) and the estimate will be of $\overline{\chi}$.

The desired comparison during fading conditions between the proposed method and estimation via the SER is achieved by comparing (170) to (174). This is done in Fig. 85 to Fig. 88. The lowest SNRs for which results are given in Fig. 85 to Fig. 88 are rough thresholds $\overline{\Gamma}_M$ defined as the average SNRs at which $\overline{g}_M(\overline{\Gamma}_M) = 5 \cdot 10^{-2}$.

It should be noted that due to the fact that the SER often requires orders-of-magnitude more symbols, when fading is present and estimation via the SER is attempted, it is likely that we will have $L \gg T_{COH}$ whereas we would still have $2NT \ll T_{COH}$ for estimation SNR estimation via $\hat{l}_{M,N}$. In such a case, the correct comparison would be curves for case (a) ($2NT \ll T_{COH}$) vs. curves from case (b) ($L \gg T_{COH}$). Such a comparison can easily be done by looking at the appropriate curves from Fig. 85 to Fig. 88 and Fig. 69-Fig. 71.

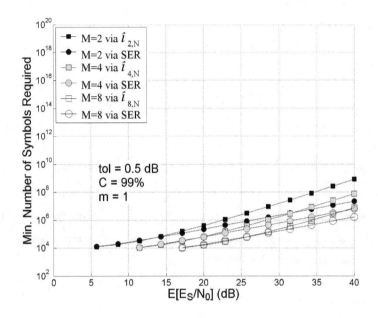

Fig. 85. Estimation via $\hat{l}_{M,N}$ vs. estimation via the SER, for case (b) ($2NT \gg T_{COH}$) and case (ii) ($L \gg T_{COH}$) for Nakagami-*m* fading with $m=1$.

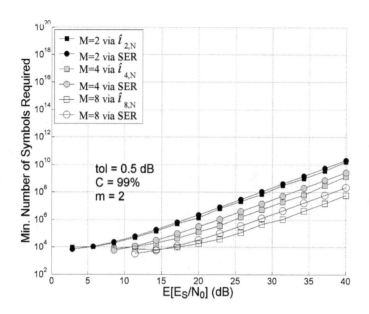

Fig. 86. Estimation via $\hat{l}_{M,N}$ vs. estimation via the SER, for case (b) ($2NT \gg T_{COH}$) and case (ii) ($L \gg T_{COH}$) for Nakagami-m fading with $m = 2$.

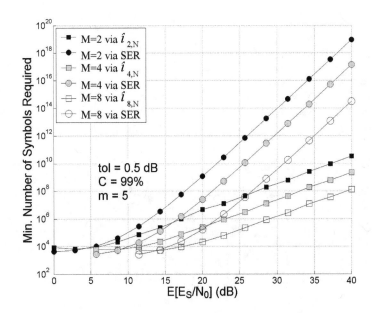

Fig. 87. Estimation via $\hat{l}_{M,N}$ vs. estimation via the SER, for case (b) ($2NT \gg T_{COH}$) and case (ii) ($L \gg T_{COH}$) for Nakagami-m fading with $m = 5$.

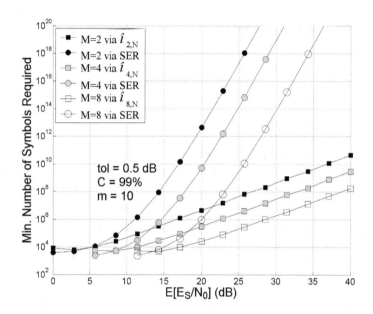

Fig. 88. Estimation via $\hat{l}_{M,N}$ vs. estimation via the SER, for case (b) ($2NT \gg T_{COH}$) and case (ii) ($L \gg T_{COH}$) for Nakagami-m fading with $m=10$.

4.7.3   Discussion

As can be seen from inspection of Fig. 85 to Fig. 88, estimation of the average SNR via $\hat{l}_{M,N}$ often requires orders of magnitude less symbols as compared to estimation via the SER. Looking first at high SNR, we note that estimation via $\hat{l}_{M,N}$ generally requires much less symbol intervals than estimation via the SER. Thus, by estimating from $\hat{l}_{M,N}$ we can generate estimates much more rapidly than by estimating the SNR from the SER. We also see that as the fading index m increases, so does the advantage of the proposed estimator. For moderate and high m, and, as well, for case (a) ($2NT \ll T_{COH}$) we find that

- 181 -

estimation via $\hat{l}_{M,N}$ is much better than estimation via the SER, often by many orders of magnitude.

At low SNRs, we see that estimation via $\hat{l}_{M,N}$ requires about the same number of symbols as estimation via the SER. However, consider the case of unknown data being transmitted. The sole way to obtain an SER estimate from unknown data is by via code-decode process [62]. Specifically, the transmitted data is coded and then decoded at the receiver (e.g., using block codes or convolution codes [13 Chap. 8]), an error rate estimate the receiver is obtained by comparing the decoded data stream to the input data stream. However, this implicitly assumes that the error correction decoder's output is completely error free, which is a bad assumption at low SNRs. Thus, the SER estimate would be inherently more unreliable as the SNR decreases, and, moreover, the error correction decoder (ECD) might not even be locked at low SNRs. As one method of countering this effect, training sequences, pilot symbols, or preambles can be sent over the channel and the SER estimation can be done upon those known symbols. However, this incurs a reduction in the channel's information-bearing capacity, since channel throughput that is taken up by those known symbols cannot be used in order to transmit data. Secondly, if we call the percentage of known symbols in the data stream P (e.g., P=10%), then we have that the number of symbol intervals that we actually have to wait in order to arrive at the SER estimate is increased by a factor of $1/P$ over the quantities predicted by Fig. 85 to Fig. 88, i.e. for P=10% we need to multiply those quantities by 10, which clearly degrades the performance of estimation via the SER as compared to estimation via $\hat{l}_{M,N}$.

In conclusion, due to the aforementioned reasons Fig. 85 to Fig. 88 present optimistic results with regards to SNR estimation via the SER. Therefore, the proposed method is superior (in terms of estimation latency) at low SNRs as well.

In terms of hardware complexity, estimation via the SER requires the implementation of error detection and accrual mechanisms. This often means the allocation of sizeable hardware and/or software resources for this purpose. Moreover, a lookup table that would translate the SER measurement into an SNR measurement is still required, and such a lookup table is much larger that the one presented in Sec. 4.3.2 (see discussion in Sec. 4.5.5) .

4.8 SNR Estimation via $\hat{l}_{M,N}$ for M-PSK in the Presence of Fading: NMSE evaluation

When fading is present, the SNR estimation approach taken in recent years has been to estimate the SNR using Viterbi algorithm-based [15] or EM (Expectation Maximization) algorithm-based estimators [70]. Obviously, such estimators have many orders-of magnitude more complexity than the proposed estimator. As for previously studied blind 1 sample/symbol moment-based estimation, it has been documented that the $M_2 M_4$ performs very poorly in fading conditions [70].

In order to evaluate the effects of fading upon the proposed estimator's NMSE, simulation results are obtained for the case of Nakagami-m fading for various values of *m*, and the NMSE is compared to the NMSE without fading.

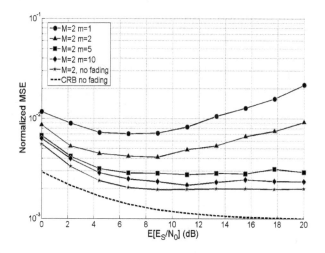

Fig. 89. NMSE comparison of SNR estimation via estimation via $\hat{l}_{M,N}$ for BPSK and Nakagami-m fading. For each metric $2N = 1024$ symbols were used.

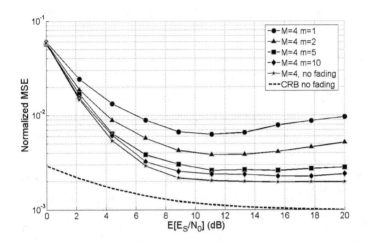

Fig. 90. NMSE comparison of SNR estimation via estimation via $\hat{l}_{M,N}$ for QPSK and Nakagami-m fading. For each metric $2N = 1024$ symbols were used.

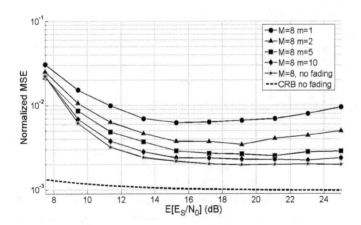

Fig. 91. NMSE comparison of SNR estimation via $\hat{l}_{M,N}$ for 8-PSK and Nakagami-m fading. For each metric $2N = 1024$ symbols were used.

Looking at Fig. 89-Fig. 91, we see that over all the SNR range of interest, the SNR estimator performs quite well, that is, its NMSE is below $5 \cdot 10^{-2}$ and usually below 10^{-2}, which is quite a useful range (see Sec. 4.6.2). An increase in the NMSE is observed at high-SNR, although this is not particularly problematic since SNR estimates need not be very accurate at high SNRs (since palpable performance gains by employing SNR estimates will be achieved only in low and moderate SNRs). Furthermore, it should be noted that the NMSE can be reduced simply by increasing the number of symbols $2N$ that is used in order to compute the estimator. This will have a negligible effect on the estimators complexity: the only increase in complexity is the augmentation of the accumulator register and adder in the IAD structure by a few bits - see Fig. 24 (for example, if we increase $2N$ to $2N = 1024 \cdot 16 = 16384$, we would need to augment these structures by $\log_2 16 = 4$ bits). Hence, the degradation due to fading can be easily overcome by increasing the estimation period and with a negligible increase in the estimator complexity.

Now, let us look at the effect of fading on the estimator's bias. This is shown in the following figures.

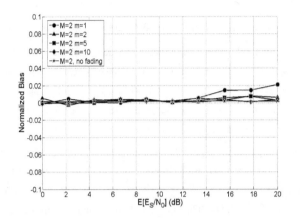

Fig. 92. Normalized bias comparisons for SNR estimation via $\hat{l}_{M,N}$ for BPSK and Nakagami-m fading. For each metric $2N = 1024$ symbols were used.

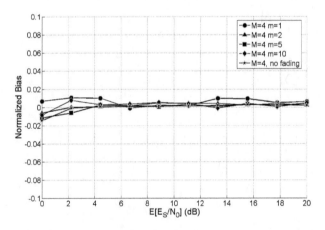

Fig. 93. Normalized bias comparisons for SNR estimation via $\hat{l}_{M,N}$ for QPSK and Nakagami-m fading. For each metric $2N = 1024$ symbols were used.

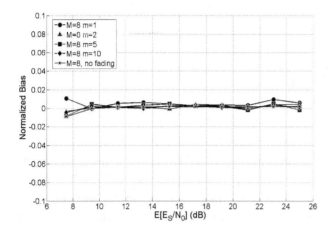

Fig. 94. Normalized bias comparisons for SNR estimation via $\hat{l}_{M,N}$ for 8-PSK and Nakagami-m fading. For each metric $2N = 1024$ symbols were used.

As can be seen in Fig. 92-Fig. 94, there is no appreciable effect of fading on the estimator's bias and it remains very near 0 for all SNRs of interest.

To conclude this section, we note that that the effects of fading on the proposed estimator mild and manageable. This is a sharp contrast to the performance of the $M_2 M_4$ estimator [70]. Moreover, the complexity of the proposed estimator (which is NDA and operates at 1 sample/symbol) is significantly lower than that of other estimators which have been proposed for SNR estimation in fading conditions [15],[70].

Part B   A New SNR Estimation Structure for D-MPSK and for M-PSK in the Absence of Carrier Synchronization.

4.9   Introduction to Part B of This Chapter

D-MPSK (Differential M-ary Phase Shift Keying) systems differ from M-PSK systems in that in a D-MPSK receiver there is no carrier synchronization loop, but, rather, the demodulation is done by differential detection (see, for example, [13 Sec. 5.2.8], [109 Chap. 10]). In D-MPSK, like for M-PSK, one of the most important signal metrics in any receiver's operation is an estimate of the received signal's SNR. Some applications of such a metric can be found in Sec. 4.1 and will not be repeated here.

In this, Part B of Chapter 4, we present a robust real-time SNR estimator for D-MPSK. This estimator is a modification of the estimator for coherent M-PSK presented in Part A of this chapter, and, as such, it retains the advantages which were observed for that estimator. Specifically, the estimator is shown to have the following advantages:

(i)   The estimator has a compact fixed-point hardware implementation which is quite suitable for implementation within FPGAs or ASICs.

(ii)   The estimator requires only 1 sample/symbol.

(iii)   Accurate estimates can be generated in real-time.

(iv)   The estimator is resistant to imperfections in the AGC (Automatic Gain Control) circuit.

We investigate the proposed estimator theoretically and through simulations. General formulas are developed for SNR estimation in the presence of frequency-flat slow fading (i.e. where $T_{COH} \ll T$ and $1/T \ll W_{COH}$), and specific results are presented for Nakagami-m fading. The proposed estimator is then compared to other SNR estimators, and it is shown that the proposed method requires less hardware resources while at the same time providing superior performance. Finally, we consider application of the proposed D-MPSK SNR estimator to SNR estimation in coherent M-PSK receivers when carrier synchronization has not been achieved. The estimator is shown to have excellent performance and an exceptionally compact hardware implementation.

Organization of this part of the chapter is as follows. In Section 4.10 we briefly outline the system model upon which the discussion is undertaken. In Section 4.11 we present the motivation for the new SNR estimator as well as its hardware implementation. Stochastic analysis of the SNR estimation process is pursued in 4.12 and 4.13. In Sections 4.14 and 4.15 we compare the efficacy of the proposed estimator to other estimators. In Section 4.16 we discuss the application the SNR estimator to M-PSK systems where the carrier PLL is unlocked. Finally, Section 4.17 is devoted to conclusions.

4.10 D-MPSK System Model

Signal and receiver characteristics are assumed identical to those of Part A, except that here (since we assume D-MPSK demodulation) we do not have a carrier recovery PLL. (Alternatively, as we discuss in Section 4.16, the system may be considered identical to that of Part A, except that the carrier PLL is assumed to be unlocked (with certain conditions upon $\Delta\omega$, do be established below)) . The reader is strongly urged to take a thorough look at Part A of this chapter and Section 1.4, since notations and results given there will be used here extensively.

The baseband PSK signal is $m(t) \triangleq \sum_{n=-\infty}^{\infty} a_n p(t-nT)$, with $p(t)$ being the pulse shape and $a_n = \exp(j\phi_n)$, $\phi_n = 2\pi \cdot m_n / M$, with $m_n \in \{0,1,...,M-1\}$. The modulated signal is $s_m(t) \triangleq \mathrm{Re}[m(t)\exp(j\omega_i t + j\theta_i)]$. A simplified diagram of the front-end of the receiver under discussion is shown in Fig. 95. At the receiver, from Section 1.4 we have

$$I(n) = K\left(2E_S \cdot \cos\left(-\Delta\omega \cdot nT + \theta_e + \phi_n\right) + n_I(nT)\right) \qquad \text{and}$$

$$Q(n) = K\left(2E_S \cdot \sin\left(-\Delta\omega \cdot nT + \theta_e + \phi_n\right) + n_Q(nT)\right), \qquad \text{with} \qquad \theta_e \triangleq \theta_i - \theta_o \qquad \text{and}$$

$n_I(nT), n_Q(nT) \sim N(0, 2N_0 E_S)$. K is the equivalent (AGC-controlled) I-Q arm gain (see Sec. 1.5 for a thorough discussion of the AGC and the parameter K). The complex symbol is:

$$r_n \triangleq I(n) + j \cdot Q(n) \qquad (175)$$

and its phase is:

- 189 -

$$\varphi_n \triangleq \tan^{-1}\left(Q(n)/I(n)\right) \tag{176}$$

We then have:

$$r_n = |r_n| \exp\left(j\varphi_n\right) \tag{177}$$

Here (unlike in Part A), we do not assume $\Delta\omega=0$, but rather $|\Delta\omega| \ll 2\pi/(M\cdot T)$. It should be noted that the assumption $|\Delta\omega| \ll 2\pi/(M\cdot T)$ is the standard assumption that is made in D-MPSK receivers (see for example [27 Sec. 10.19]).

We assume that our signal is subject to a frequency-flat (= frequency-nonselective) channel with slow fading (i.e. where $T_{COH} \ll T$ and $1/T \ll W_{COH}$. For the definition of such a process, see for example [13 Sec. 14.3]). We shall comment that, in general, D-MPSK would perform poorly in frequency-selective or fast fading (in which a different, probably noncoherent, multicarrier, or spread-spectrum type modulation would be chosen see [13 p. 818], [78], [79]). Thus, the assumption of frequency-flat slow fading will be appropriate for most D-MPSK systems.

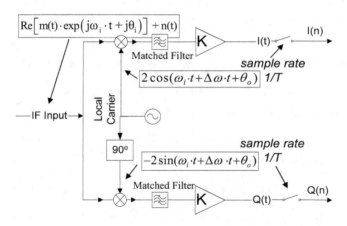

Fig. 95. Front end of D-MPSK receiver (simplified diagram).

4.11 Motivation and Estimator Structure

4.11.1 Motivation

Detection of D-MPSK signals is often facilitated by first generating a *pseudo-coherently demodulated* M-PSK signal

$$u_n \triangleq r_n r_{n-1}^* \tag{178}$$

and then applying M-PSK decision regions upon u_n. The motivation here is similar, but we add a twist: the idea is to use the estimator of Part A upon a *normalized* pseudo-coherently demodulated M-PSK signal

$$v_n \triangleq \frac{r_n r_{n-1}^*}{|r_n||r_{n-1}|} \tag{179}$$

As we shall see, using v_n instead of u_n yields a simpler hardware implementation.

4.11.2 Estimator structure and operation principle

We define:

$$x_{M,n}^D \triangleq \mathrm{Re}[(v_n)^M]/|v_n|^M \tag{180}$$

(Note: to avoid confusion with Part A, throughout this part of the chapter we use superscript "D" in variables pertaining to D-MPSK structures). The estimator of Part A applied to v_n is defined as:

$$\hat{l}_{M,N}^D \triangleq \frac{1}{2N} \sum_{n=-N+1}^{N} x_{M,n}^D = \frac{1}{2N} \sum_{n=-N+1}^{N} \mathrm{Re}[(v_n)^M]/|v_n|^M \tag{181}$$

Here we do not use $\hat{l}_{M,N}^D$ as a lock detector (since there is no carrier PLL) but rather only as an SNR estimator. Note that $|v_n|=1$ for all n, so theoretically we could have defined $\hat{l}_{M,N}^D \triangleq \frac{1}{2N} \sum_{n=-N+1}^{N} \mathrm{Re}[(v_n)^M]$. However, when quantization effects are taken into account we see that $|v_n|=1$ does not always hold, and then $\hat{l}_{M,N}^D \triangleq \frac{1}{2N} \sum_{n=-N+1}^{N} \mathrm{Re}[(v_n)^M]/|v_n|^M$ has distinct implementation and performance advantages (which are outlined in Section 4.11.3).

In this part of the chapter we shall present a general method for D-MPSK SNR estimation in the presence of fading. As noted earlier, we assume that the fading is slow (i.e. $T_{COH} \gg T$, where T_{COH} is the channel coherence time) and that the channel is frequency-nonselective. Once again we use the notation χ to refer to the instantaneous SNR, and the notation $\overline{\chi}$ to denote the *average* SNR (i.e., $\overline{\chi} \triangleq E[E_S / N_0]$). The conditional pdf (probability density function) of the SNR due to fading will be denoted as $p_F(\chi|\overline{\chi}) \triangleq p(E_S / N_0 = \chi | E[E_S / N_0] = \overline{\chi})$. For example, from [77 Table 2] we have for

Rayleigh fading $\quad p_F(\chi|\overline{\chi}) = \frac{1}{\overline{\chi}} \exp\left(\frac{-\chi}{\overline{\chi}}\right) \quad$ and for Nakagami-m fading

$p_F(\chi|\overline{\chi}) = \frac{m^m \chi^{m-1}}{\overline{\chi}^m \Gamma(m)} \exp\left(\frac{-m\chi}{\overline{\chi}}\right)$. Other common fading distributions were given in Table 2.

We differentiate between two cases:

(a) $2NT \ll T_{COH}$

(b) $2NT \gg T_{COH}$

For case (a), during the averaging over $2NT$ symbol intervals that is done in eq. (181) the channel SNR will not have changed much; hence SNR estimation from $\hat{l}_{M,N}^D$ will yield an estimate of the *instantaneous* SNR ratio χ.

For case (b), since $2NT$ is much larger than the channel coherence time, the distribution of SNR values encountered during the estimator computation will follow $p_F(\chi|\overline{\chi})$, and SNR estimation from $\hat{l}_{M,N}^D$ will yield an estimate of the *average* SNR ratio $\overline{\chi}$.

As already noted in Part A of this chapter, in general we would ideally like to produce SNR estimates that are instantly available and can be fed in real-time to the decoder, equalizer, or other receiver components which could make good use of them. Hence, ideally, we would be served by perfect knowledge of the instantaneous SNR ratio χ (which, if desired, could be averaged over time in order to produce an estimate of $\overline{\chi}$). However, estimation of χ is not always possible, due to the fact that T_{COH} may be too short as compared to the estimation period $2NT$ which is necessary in order to achieve an acceptable accuracy in the SNR estimation (see Secs. 4.13-4.14). Nonetheless, if $2NT \gg T_{COH}$, timely knowledge of the average SNR $\bar{\chi}$ is often sufficient in order to facilitate substantial performance gains [15].

Estimation is achieved following a procedure analogous to (113), i.e. we estimate the SNR via the following:

Case (a) ($2NT \ll T_{COH}$) : the instantaneous SNR is estimated through:

$$\gamma_{dB}^{D} = 10 \cdot \log_{10} \left(\left(f_{M}^{D} \right)^{-1} (\hat{i}_{M,N}^{D}) \right) \tag{182}$$

where $f_{M}^{D}(\chi) \triangleq E\left[\hat{i}_{M,N}^{D} \middle| E_S / N_0 = \chi \right]$

Case (b) ($2NT \gg T_{COH}$) : the average SNR is estimated through:

$$\bar{\gamma}_{dB}^{D} = 10 \cdot \log_{10} \left(\left(\bar{f}_{M}^{D} \right)^{-1} (\hat{i}_{M,N}^{D}) \right) \tag{183}$$

where $\bar{f}_{M}^{D}(\bar{\chi}) \triangleq E\left[\hat{i}_{M,N}^{D} \middle| E[E_S / N_0] = \bar{\chi} \right]$

4.11.3   Hardware implementation

A fixed-point (2's complement) hardware implementation of the estimator is shown in Fig. 96. The LUTs (Lookup Tables) require

$$nbits = \underset{\underset{LUT\,\#1}{\uparrow}}{2b_2 2^{2b_1}} + \underset{\underset{LUT\,\#2}{\uparrow}}{2b_2 2^{2b_1}} + \underset{\underset{LUT\,\#3}{\uparrow}}{b_4 2^{2b_3}} + \underset{\underset{LUT\,\#4}{\uparrow}}{b_5 2^{b_4}} \tag{184}$$

bits. See Secs. 2.3.3 and 4.3.2 for discussions applicable to LUT #3 and LUT #4, as well as discussion of why $2N$ should be a power of 2. The use of v_n rather than u_n significantly reduces the hardware resources needed to compute $\hat{i}_{M,N}^{D}$; this is because the normalized constellation has less dynamic range (it is $[-1,1]$) so this reduces b_2 and b_3 required to achieve an acceptable degradation due to quantization, as we shall see in Sec. 4.12.

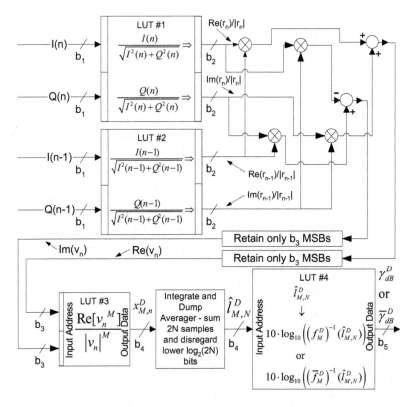

Fig. 96. Fixed-point hardware generation of γ_{dB}^D.

4.12 Conditional Distribution of $\hat{\imath}_{M,N}^D$

In this section we shall derive the conditional probability distribution of $\hat{\imath}_{M,N}^D$. These stochastic properties will then be used to develop the SNR estimation method in Section 4.13.

For simplicity and without loss of generality (see Sec. 2.4) we assume $\forall n, \phi_n = 0$, whereupon from (176):

$$\varphi_n = \tan^{-1}\left(\frac{\sin(-\Delta\omega \cdot nT + \theta_e) + n_Q(nT)/(2E_S)}{\cos(-\Delta\omega \cdot nT + \theta_e) + n_I(nT)/(2E_S)}\right) \tag{185}$$

Let us define (similar to (43)):

$$\Delta\phi_n \triangleq \varphi_n - \left(-\Delta\omega nT + \theta_e\right) \tag{186}$$

Since we assumed $\forall n, \phi_n = 0$ then the physical meaning of $\Delta\phi_n$ is clear: it is the phase error in the received symbol that can be attributed to $n_I(nT)$ and $n_Q(nT)$ (to see this, substitute $n_I(nT) = n_Q(nT) = 0$ in the expressions for $I(n)$ and $Q(n)$, and then $\varphi_n = \tan^{-1}\left(Q(n)/I(n)\right) = -\Delta\omega nT + \theta_e \Rightarrow \Delta\phi_n = \varphi_n - (-\Delta\omega nT + \theta_e) = 0$).

Since (185) has the same form as (42) with θ_e replaced by $-\Delta\omega nT + \theta_e$, we have that $\Delta\phi_n$ as defined in (186) is distributed the same as $\Delta\phi_n$ as defined in (43), namely at $E_S / N_0 = \chi$ it has a Rician phase pdf (probability density function) given by (29):

$$p_R\left(\Delta\phi|\chi\right) \triangleq p\left(\Delta\phi_n = \Delta\phi | E_S / N_0 = \chi\right)$$

$$= \frac{\exp(-\chi)}{2\pi}\left[1 + \sqrt{2\chi}\cos(\Delta\phi)\exp\left(\chi\cdot\cos^2(\Delta\phi)\right)\cdot\int_{-\infty}^{\cos(\Delta\phi)\sqrt{2\chi}} e^{-y^2/2}dy\right] \tag{187}$$

where $-\pi \le \Delta\phi \le \pi$.

Now, let us investigate v_n. Trivial substitutions of (177) and (178) into (179) show that

$$v_n = \frac{|r_n|\exp(j\varphi_n)|r_{n-1}|\exp(-j\varphi_{n-1})}{|r_n||r_{n-1}|} = \exp\left(j\varphi_n^D\right) \tag{188}$$

where

$$\varphi_n^D \triangleq \varphi_n - \varphi_{n-1} \tag{189}$$

Observe that $\varphi_n^D \in [-2\pi, 2\pi]$, though the true phase is $\varphi_n^D \bmod_{2\pi} \in [-\pi, \pi]$. We could have indeed performed the modulo operation and confined the range of φ_n^D to $[-\pi, \pi]$; this is, in fact, the approach undertaken in [110], [111], and [112]. In contrast, we choose to follow the approach of [113] and to maintain the pretence $\varphi_n^D \in [-2\pi, 2\pi]$ because, as we shall see, it simplifies the analysis. However, note that since $v_n = \exp\left(j\varphi_n^D\right) = \exp\left(j\left(\varphi_n^D \bmod_{2\pi}\right)\right)$ this choice has no bearing upon the results. From (186) we then have:

$$\varphi_n^D = \varphi_n - \varphi_{n-1} = \Delta\phi_n + \left(-\Delta\omega \cdot nT + \theta_e\right) - \left(\Delta\phi_{n-1} + \left(-\Delta\omega \cdot (n-1)T + \theta_e\right)\right)$$
$$= \Delta\phi_n - \Delta\phi_{n-1} - \Delta\omega T \tag{190}$$

Let us define $\Delta\phi_n^D \triangleq \Delta\phi_n - \Delta\phi_{n-1}$; note that since $\Delta\phi_n, \Delta\phi_{n-1} \in [-\pi, \pi]$ then $\Delta\phi_n^D \in [-2\pi, 2\pi]$. The pdf of $\Delta\phi_n^D$ is easily found since it is a convolution of the distributions of $\Delta\phi_n$ and $\left(-\Delta\phi_{n-1}\right)$, namely (for $-2\pi \le \Delta\phi^D \le 2\pi$)

$$p_D\left(\Delta\phi^D \middle| \chi\right) \triangleq p\left(\Delta\phi_n^D = \Delta\phi^D \middle| E_S / N_0 = \chi\right) = \int_{-\pi}^{\pi} p_R(\tau) p_R\left(\tau - \Delta\phi^D\right) d\tau \tag{191}$$

which is straightforward to evaluate numerically. We note that though closed form expressions for (191) seem unattainable, a Fourier series representation for the pdf is given in [113 eq. (4)] (substituting $\theta = 0$ there)). Moreover, a simple expression for the distribution at high SNR is easily obtained: from (34) we have $\forall n, \Delta\phi_n \overset{\chi \to \infty}{\sim} N\left(0, 1/(2\chi)\right)$; now, since $\Delta\phi_n^D = \Delta\phi_n - \Delta\phi_{n-1}$ and since $\Delta\phi_n$ and $\Delta\phi_{n-1}$ are independent (see Sec. 2.3) we have $\Delta\phi_n^D \overset{\chi \to \infty}{\sim} N\left(0, 1/\chi\right)$, i.e.

$$p_D\left(\Delta\phi^D \middle| \chi\right) \overset{\chi \to \infty}{\approx} \sqrt{\chi/2\pi} \cdot \exp\left(-0.5 \cdot \chi \cdot \left(\Delta\phi^D\right)^2\right) \tag{192}$$

We now have the tools necessary in order to investigate the distribution of $\hat{I}_{M,N}^D$.

4.12.1 Conditional expectation of $\hat{I}_{M,N}^D$ given χ for $2NT \ll T_{COH}$

For $2NT \ll T_{COH}$, we can assume that the E_S / N_0 ratio is constant over the estimation interval and is equal to the instantaneous SNR χ. Hence:

$$f_M^D(\chi) \triangleq E\left[\hat{i}_{M,N}^D \middle| \frac{E_S}{N_0} = \chi\right]$$

$$= E\left[\text{Re}\left[\left(\cos\varphi_n^D + j\cdot\sin\varphi_n^D\right)^M\right] \middle/ \left(\cos^2\varphi_n^D + \sin^2\varphi_n^D\right)^{M/2} \middle| \frac{E_S}{N_0} = \chi\right]$$

$$= E\left[\cos\left(M\varphi_n^D\right) \middle| \frac{E_S}{N_0} = \chi\right] = E\left[\cos\left(M\Delta\phi_n^D - M\Delta\omega T\right) \middle| \frac{E_S}{N_0} = \chi\right]$$

$$= E\left[\cos\left(M\Delta\phi_n^D\right) \middle| \frac{E_S}{N_0} = \chi\right]\cos\left(M\Delta\omega T\right) \tag{193}$$

$$+ \underbrace{E\left[\sin\left(M\Delta\phi_n^D\right) \middle| \frac{E_S}{N_0} = \chi\right]\sin\left(M\Delta\omega T\right)}_{=0}$$

$$= E\left[\cos\left(M\Delta\phi_n^D\right) \middle| \frac{E_S}{N_0} = \chi\right]\cos\left(M\Delta\omega T\right)$$

$$= \left(\int_{-2\pi}^{2\pi}\cos\left(M\Delta\phi\right)p_D\left(\Delta\phi\middle|\chi\right)d\left(\Delta\phi\right)\right)\cos\left(M\Delta\omega T\right)$$

As was the case with $f_M(\chi)$, it is possible to find closed-form expressions for $f_M^D(\chi)$. This is done in Appendix C. We find using those derivations that:

$$f_M^D(\chi) = \frac{\pi\chi}{4}\cdot e^{-\chi}\left[I_{\frac{M-1}{2}}\left(\frac{\chi}{2}\right) + I_{\frac{M+1}{2}}\left(\frac{\chi}{2}\right)\right]^2\cos\left(M\Delta\omega T\right) \tag{194}$$

which (for even M) can be simplified to (see Appendix C):

$$f_M^D(\chi) =$$

$$\left(\begin{array}{l}\dfrac{M^2}{4}\displaystyle\sum_{n=0}^{M/2}\sum_{k=0}^{M/2}\dfrac{(-1)^{n+k}}{n!k!}\cdot\dfrac{(M/2+n-1)!(M/2+k-1)!}{(M/2-n)!(M/2-k)!\chi^{n+k}}\\[3mm]
+(-1)^{M/2+1}\,M\cdot e^{-\chi}\displaystyle\sum_{n=0}^{M/2}\sum_{k=1}^{M/2}\dfrac{(-1)^n}{n!(k-1)!}\dfrac{(M/2+n-1)!(M/2+k-1)!}{(M/2-n)!(M/2-k)!\chi^{n+k}}\\[3mm]
+\exp(-2\chi)\displaystyle\sum_{n=1}^{M/2}\sum_{k=1}^{M/2}\dfrac{(M/2+n-1)!(M/2+k-1)!}{(n-1)!(k-1)!(M/2-n)!(M/2-k)!\chi^{n+k}}\end{array}\right) \tag{195}$$

$$\times\cos\left(M\Delta\omega T\right)$$

It is emphasized that the sums in (195) have a finite number of terms and can thus be easily and accurately computed.

At high SNR we use (192) to obtain a useful approximation (using $\int_0^{\infty} e^{-ax^2}\cos(bx)dx = \frac{1}{2}\sqrt{\frac{\pi}{a}}e^{-b^2/(4a)}$ [54 eq. 15.73]):

$$f_M^D(\chi) \overset{\chi\to\infty}{\approx} \left(\int_{-\infty}^{\infty} \cos(M\Delta\phi)\sqrt{\frac{\chi}{2\pi}}\exp\left(-\frac{\chi}{2}(\Delta\phi)^2\right)d(\Delta\phi)\right)\cos(M\Delta\omega T)$$

$$= \exp\left(-\frac{M^2}{2\chi}\right)\cos(M\Delta\omega T) \qquad (196)$$

We assumed (see Sec. 4.10) as is appropriate for the operating point of D-MPSK systems, that $|\Delta\omega| \ll 2\pi/(M\cdot T)$ holds, implying $\cos(M\Delta\omega T)\approx 1$. Therefore, the degradation in (193)-(196) due to carrier frequency error is negligible. We thus henceforth assume for simplicity $\Delta\omega = 0$, though we note that (193)-(196) provide an easy way to incorporate modeling of small frequency errors.

Plots of (193), (196), and simulated results for $\Delta\omega = 0$ are given in Fig. 97; we see that (196) is an excellent approximation. The simulations in Fig. 97 which include quantization effects are quite realistic since they model the following AGC effects: (a) sampler input signal-level backoff (samplers are assumed to be driven at an RMS (Root-Mean-Square) of 80% of the samplers' full-scale voltage range) and (b) clamping by the samplers when they are saturated. The AGC is assumed to behave as the example AGC described in Section 1.5. Hence, the simulations presented should be a good prediction of achievable results. If we assume for example $b_s = 8$ (which would imply an 8-bit SNR measurement[18] in dB – usually more than sufficient), then from (184) we have for the simulated quantized systems in Fig. 97 that *nbits*= 14336, 30720, 124928, respectively, all of which are very reasonable considering the amount of dedicated memory available in contemporary FPGAs (e.g. the various Xilinx Virtex families [114] or Altera Stratix families [115]) or which can be implemented in ASICs. Fig. 97 shows that for low M

---

[18] Note that such a measurement could include digits after the binary point. For example, if we put the binary point to the left of the Least Significant Bit (LSB), then we have for an 8-bit output the following: 1 sign bit followed by 6 whole-number bits and 1 fractional bit, which would allow, in 2's complement notation, the representation of the interval -64 dB to +63 dB in 0.5 dB intervals, which is usually quite sufficient range and quantization.

only coarse quantization is needed, while (as expected) higher Ms require finer quantization to achieve good agreement with the predicted value of $\hat{i}^{D}_{M,N}$.

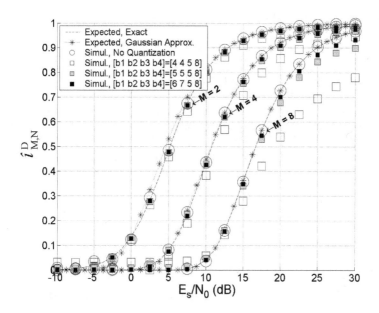

Fig. 97. Expected and simulated values of $\hat{i}^{D}_{M,N}$ for case (a).

4.12.2 Conditional expectation of $\hat{i}^{D}_{M,N}$ given $\overline{\chi}$ for $2NT \gg T_{COH}$

For the case of $2NT \gg T_{COH}$, since $2NT$ is much larger than the channel coherence time, the distribution of SNR values encountered during the estimator computation will follow $p_F\left(\chi|\overline{\chi}\right)$, and SNR estimation from $\hat{i}^{D}_{M,N}$ will yield an estimate of the *average* SNR ratio $\overline{\chi}$. The expectation of $\hat{i}^{D}_{M,N}$ conditioned upon $\overline{\chi}$ is:

$$\overline{f}_M^D(\overline{\chi}) \triangleq E\left[\hat{i}_{M,N}^D \middle| E\left[\frac{E_S}{N_0}\right] = \overline{\chi}\right]$$

$$= E\left[\operatorname{Re}\left[\left(\cos\varphi_n^D + j\cdot\sin\varphi_n^D\right)^M\right] \middle/ \left(\cos^2\varphi_n^D + \sin^2\varphi_n^D\right)^{M/2} \middle| E\left[\frac{E_S}{N_0}\right] = \overline{\chi}\right]$$

$$= E\left[\cos\left(M\varphi_n^D\right) \middle| E\left[\frac{E_S}{N_0}\right] = \overline{\chi}\right] = E\left[\cos\left(M\Delta\phi_n^D - M\Delta\omega T\right) \middle| E\left[\frac{E_S}{N_0}\right] = \overline{\chi}\right]$$

$$= E\left[\cos\left(M\Delta\phi_n^D\right) \middle| E\left[\frac{E_S}{N_0}\right] = \overline{\chi}\right]\cos\left(M\Delta\omega T\right) \qquad (197)$$

$$+ \underbrace{E\left[\sin\left(M\Delta\phi_n^D\right) \middle| E\left[\frac{E_S}{N_0}\right] = \overline{\chi}\right]\sin\left(M\Delta\omega T\right)}_{=0}$$

$$= E\left[\cos\left(M\Delta\phi_n^D\right) \middle| E\left[\frac{E_S}{N_0}\right] = \overline{\chi}\right]\cos\left(M\Delta\omega T\right)$$

$$= \left(\int_0^\infty \left(\int_{-2\pi}^{2\pi} \cos\left(M\Delta\phi^D\right)p_D\left(\Delta\phi^D \middle| \chi\right)d\left(\Delta\phi^D\right)\right)p_F\left(\chi \middle| \overline{\chi}\right)d\chi\right)\cos\left(M\Delta\omega T\right)$$

Another expression of (197) is simply through:

$$\overline{f}_M^D(\overline{\chi}) = \int_0^\infty f_M^D(\chi)p_F\left(\chi \middle| \overline{\chi}\right)d\chi \qquad (198)$$

Eq. (198) will be of particular importance in Appendix D, where closed-form expressions for $\overline{f}_M^D(\overline{\chi})$ are developed.

At high $\overline{\chi}$ we can also assume that the instantaneous SNR χ is also high, and we use (192) to obtain a useful approximation (using $\int_0^\infty e^{-ax^2}\cos(bx)dx = \frac{1}{2}\sqrt{\frac{\pi}{a}}e^{-b^2/(4a)}$ [54 eq. 15.73]):

$$\overline{f}_M^D(\overline{\chi}) \overset{\overline{\chi}\to\infty}{\approx}$$

$$\left(\int_0^\infty \int_{-\infty}^\infty \cos\left(M\Delta\phi^D\right)\sqrt{\frac{\chi}{2\pi}}\exp\left(\frac{-\chi}{2}\left(\Delta\phi^D\right)^2\right)d\left(\Delta\phi^D\right)p_F\left(\chi \middle| \overline{\chi}\right)d\chi\right)\cos\left(M\Delta\omega T\right) =$$

$$= \left(\int_0^\infty \exp\left(\frac{-M^2}{2\chi}\right)p_F\left(\chi \middle| \overline{\chi}\right)d\chi\right)\cos\left(M\Delta\omega T\right) \qquad (199)$$

Due to the infinite number of possible fading distributions, we obviously cannot present results (197) for all fading types. Rather, for more insight into the behaviour of $\hat{i}_{M,N}^D$ when $2NT \gg T_{COH}$ we shall investigate its behaviour under Nakagami-m fading, which is a fading statistic commonly found in systems which use D-MPSK ([116],

[108]). We again assume $\Delta\omega \approx 0$ (more specifically, that $|\Delta\omega| \ll 2\pi/(M \cdot T)$, see Sec. 4.10) and plot theoretical and simulated results for (197) in Fig. 98 for various types of Nakagami-m statistics. The theoretical results were derived using Appendix D. Comparing Fig. 98 to Fig. 97 we see that the effect of the Nakagami-m fading upon the curve of $\hat{i}_{M,N}^D$ is rather mild. To evaluate the effects of fading upon the quantization requirements, we plot (197) for the various quantizations used in Fig. 97. This is done in Fig. 99. As we see by comparing Fig. 99 to Fig. 97, the quantization which was sufficient for case (a) (as shown in Fig. 97) is also sufficient for case (b). Hence, there is no appreciable impact of fading upon the hardware resources required for implementation of the proposed structure.

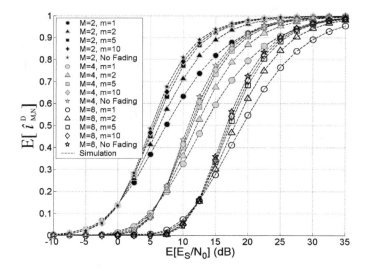

Fig. 98. $\overline{f}_M^D(\overline{\chi}) \triangleq E\left[\hat{i}_{M,N}^D \middle| E[E_S/N_0] = \overline{\chi}\right]$ as a function of $\overline{\chi}$ for Nakagami-m fading, obtained theoretically (via (254)) and via simulations, for various values of m. Quantization effects are ignored.

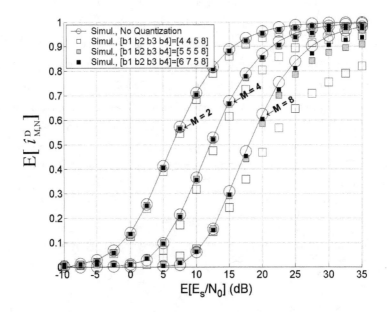

Fig. 99. Demonstration of the effects of quantization on the measured value of $\bar{f}_M^D(\bar{\chi}) \triangleq E\left[\hat{i}_{M,N}^D \Big| E[E_s / N_0] = \bar{\chi}\right]$ for Nakagami-m fading with $m=2$.

4.12.3 Variance of $\hat{i}_{M,N}^D$

It can be shown (see Appendix E) that for $|\Delta\omega| \ll 1/(M \cdot T)$ and for slow fading the cross-correlation coefficients of $\left\{x_{M,n}^D\right\}_{n=-\infty}^{\infty}$ defined as

$$\rho_{n,k} \triangleq \frac{E[x_{M,n}^D x_{M,k}^D] - E[x_{M,n}^D]E[x_{M,k}^D]}{\sqrt{\text{var}(x_{M,n}^D)\,\text{var}(x_{M,k}^D)}} \tag{200}$$

satisfy

$$\rho_{n,k} = \begin{cases} 1 & n=k \\ \rho_1(\chi) & |n-k|=1 \\ 0 & |n-k|>1 \end{cases} \tag{201}$$

where $|\rho_1(\chi)| \le 0.3$. Moreover, we can still use the derivations of Sec. 2.5 to surmise that $\forall n, \sigma_x^2 \triangleq \mathrm{var}\left(x_{M,n}^D\right) \le 1$. Now,

$$\mathrm{var}\left(\hat{l}_{M,N}^D\right) = \mathrm{var}\left(\frac{1}{2N}\sum_{n=-N+1}^{N} x_{M,n}^D\right) = \frac{1}{4N^2}\left(2N\sigma_x^2 + 2\cdot(2N-1)\rho_1(\chi)\sigma_x^2\right) \quad (202)$$

and using $\sigma_x^2 \le 1$ and $|\rho_1(\chi)| \le 0.3$ we have

$$\mathrm{var}\left(\hat{l}_{M,N}^D\right) \le \frac{1}{4N^2}\left(2N + 2\cdot(2N-1)0.3\right) \xrightarrow{\text{large } N} \frac{1.6}{2N} \quad (203)$$

Finally, from the central limit theorem for m-dependent variables [117 Chap. 7] we have that $\hat{l}_{M,N}^D$ is Gaussian.

4.12.4   Summary: Conditional distribution of $\hat{l}_{M,N}^D$

Let us unite what we have learned in the previous subsections. We have :

- For case (a) (that is, $2NT \ll T_{COH}$): $\hat{l}_{M,N}^D \sim N\left(f_M^D(\chi), \sigma^2\right)$ where $f_M^D(\chi)$ is given in (193)-(196) and $\sigma^2 \le 1.6/2N$

- For case (b) (that is, $2NT \gg T_{COH}$): $\hat{l}_{M,N}^D \sim N\left(\overline{f}_M^D(\overline{\chi}), \sigma^2\right)$ where $\overline{f}_M^D(\overline{\chi})$ is given in (197)-(199) and $\sigma^2 \le 1.6/2N$.

4.13   Principle of SNR Estimation from $\hat{l}_{M,N}^D$

As noted in Section 4.11.2, for case (a) the instantaneous SNR is estimated through $\gamma_{dB}^D = 10 \cdot \log_{10}\left(\left(f_M^D\right)^{-1}(\hat{l}_{M,N}^D)\right)$, while for case (b) the average SNR is estimated through $\overline{\gamma}_{dB}^D = 10 \cdot \log_{10}\left(\left(\overline{f}_M^D\right)^{-1}(\hat{l}_{M,N}^D)\right)$.   Graphs   of $\gamma_{dB}^D = 10 \cdot \log_{10}\left(\left(f_M^D\right)^{-1}(\hat{l}_{M,N}^D)\right)$ and $\overline{\gamma}_{dB}^D = 10 \cdot \log_{10}\left(\left(\overline{f}_M^D\right)^{-1}(\hat{l}_{M,N}^D)\right)$ are shown in Fig. 100. These curves are the value of LUT #4 in Fig. 96, and the curve to use would be chosen according to the fading characteristics of the channel. There is an additional small point that needs to be addressed: theoretically we can encounter negative values of $\hat{i}_{M,N}^D$, in which case $\overline{\gamma}_{dB}^D$ and γ_{dB}^D would be undefined. This is solved by setting the output[19] of LUT #4 to -2^{b_3-1} for $\hat{i}_{M,N}^D \leq 0$ (not shown in Fig. 100). This correctly reflects the SNR estimate for $\hat{i}_{M,N}^D \leq 0$ (which should be $-\infty$ dB, i.e. no signal) within the limits of the available quantization.

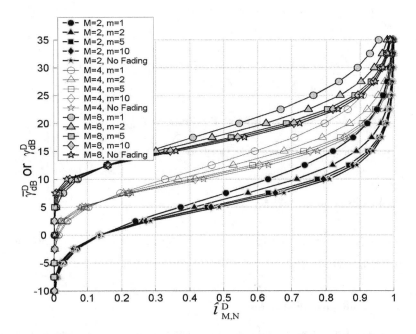

Fig. 100. γ_{dB}^D (for case (a)) or $\overline{\gamma}_{dB}^D$ (for case (b)) vs. $\hat{i}_{M,N}^D$ (the "No Fading" curves are those for case (a), while the others are for case (b)).

There is a very strong relationship between Fig. 100 and Fig. 98. To see this, recall that to graph the inverse of any monotonic function, all one has to do is reflect the graph

---

[19] This is the value of the LUT output if the representation is of whole numbers (not necessarily the case, see Footnote 18). Generally, the idea is to use the lowest SNR expressible via the LUT's quantization.

over the line $y = x$. Thus, if we reflect the curves of Fig. 98 over the line $y = x$ then we arrive at Fig. 100.

4.14 Comparison of Estimation via $\hat{i}_{M,N}^D$ to Estimation via the SER (with and without fading)

4.14.1 Number of symbols needed for estimation via $\hat{i}_{M,N}^D$

To quantitatively measure the efficacy of the SNR estimator, following Sec. 4.4 we ask:

What is the minimal value of estimation symbol intervals (which is $2 \cdot N$) needed to ensure $P\left(\left|\gamma_{dB}^D - \chi_{dB}\right| < tol\right) > C$ (for case (a)) or $P\left(\left|\overline{\gamma}_{dB}^D - \overline{\chi}_{dB}\right| < tol\right) > C$ (for case (b)) where *tol* is the tolerance and C is the confidence. For example, some appropriate values of *tol* and C would be $tol = 1.5$ dB and $C = 95\%$. Straightforward following of the development of (124) shows that the answer for case (a) is:

$$2N > 2 \cdot 1.6 \cdot \frac{\left(erf^{-1}(C)\right)^2}{\left(\min\left\{\left|f_M^D\left((1-w_2)\cdot\chi\right) - f_M^D(\chi)\right|, \left|f_M^D\left((1+w_1)\cdot\chi\right) - f_M^D(\chi)\right|\right\}\right)^2} \quad (204)$$

where $w_1 \triangleq \left(10^{tol/10} - 1\right)$ and $w_2 \triangleq \left(1 - 10^{-tol/10}\right)$. For case (b), the answer is:

$$2N > 2 \cdot 1.6 \cdot \frac{\left(erf^{-1}(C)\right)^2}{\left(\min\left\{\left|\overline{f}_M^D\left((1-w_2)\overline{\chi}\right) - \overline{f}_M^D(\overline{\chi})\right|, \left|\overline{f}_M^D\left((1+w_1)\overline{\chi}\right) - \overline{f}_M^D(\overline{\chi})\right|\right\}\right)^2} \quad (205)$$

We shall now find a comparison yardstick to which the results computed through (204) and (205) can be compared.

4.14.2 Number of symbols needed for SNR estimation via the *SER*

Following the discussion in Sec. 4.5.2 we note that many demodulators generate SNR estimates by measuring the pre- or post-decoder error rate. For example, this is done in systems that estimate the SNR from the number of errors detected in preambles, pilot symbols, or "training sequences" that are embedded in the data stream. Therefore, as noted in Sec. 4.5.2, perhaps the most meaningful and universally applicable yardstick by which to measure the efficacy of SNR estimation via $\hat{i}_{M,N}^D$ is through comparison of (204) and (205) to the number of symbols needed for SNR estimation through measurement of the pre-decoder Symbol Error Rate (SER).

SNR estimation from the SER is based upon the principle that the SER is always a strictly monotonically decreasing function of the SNR (in words: the higher the SNR, the lower the SER). Suppose we denote the SER function as $h(\chi)$ or $h(\overline{\chi})$, then if we measure the SER upon L received symbols, and denote this measured SER as $S(L)$, then an estimate of the SNR may be obtained via $h^{-1}(S(L))$. See Sec. 4.5.2 for a more thorough discussion of this point.

To address the effects of fading we must differentiate between two cases:

> **Case (i)**: $L \cdot T \ll T_{COH}$: during the counting of errors encountered during the preceding L received symbols, i.e. during the computation of $S(L)$, the channel SNR will not have changed much; hence SNR estimation from the SER will yield an estimate of the *instantaneous* SNR ratio χ.

> **Case (ii)**: $L \cdot T \gg T_{COH}$: since L is much larger than the channel coherence time, the distribution of SNR values encountered during the computation of $S(L)$ will follow $p_F(\chi|\overline{\chi})$, and SNR estimation from $S(L)$ will yield an estimate of the *average* SNR ratio $\overline{\chi}$.

The symbol error rate for D-MPSK for case (i) is [118 eq. (3)]:

$$g_M^D(\chi) \triangleq \frac{1}{\pi} \int_0^{\pi - \pi/M} \exp\left(-\chi \cdot \frac{\sin^2(\pi/M)}{1 + \cos(\pi/M)\cos\xi} \right) d\xi \tag{206}$$

The SER for D-MPSK when $L \cdot T \gg T_{COH}$ is accordingly:

$$\overline{g}_M^D(\overline{\chi}) \triangleq \int_0^\infty \left(\frac{1}{\pi} \int_0^{\pi - \pi/M} \exp\left(-\chi \cdot \frac{\sin^2(\pi/M)}{1 + \cos(\pi/M)\cos\xi} \right) d\xi \right) p_F(\chi|\overline{\chi}) d\chi \tag{207}$$

For Nakagami-m fading an even simpler formula exists for $\overline{g}_M^D(\overline{\chi})$ in the form of an integral with finite limits we have (see [108 eqs. (3), (13)], and also [116]):

$$\overline{g}_M^D(\overline{\chi}) \triangleq \frac{1}{\pi} \int_0^{\pi-\pi/M} \left(1 + \frac{\sin^2(\pi/M)}{1+\cos(\pi/M)\cos\xi} \cdot \frac{\overline{\chi}}{m}\right)^{-m} d\xi \qquad (208)$$

where the above equation was derived using the Moment Generating Function (MGF) method of SER computation (see [116],[108]). It is noted the MGF-oriented method is quite useful for computer calculations, as it involves computing only a single integral upon finite limits whose integrand is composed entirely of elementary functions.

Graphs of $g_M^D(\chi)$ (for all fading types) and $\overline{g}_M^D(\overline{\chi})$ (for Nakagami-*m* fading) are given in Fig. 101 to Fig. 105.

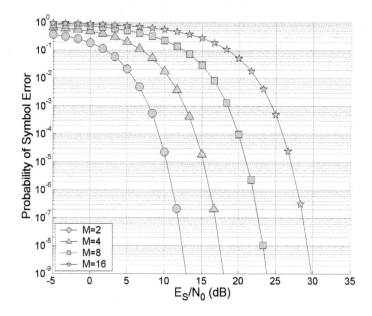

Fig. 101. Probability of symbol error $P_e = g_M^D(E_s/N_0)$ **as a function of the** E_s/N_0 **ratio.**

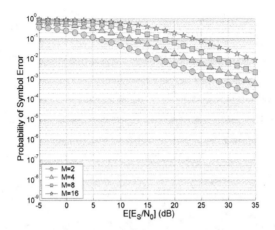

Fig. 102. Probability of symbol error $P_e = \overline{g}_M^D(\overline{\chi})$ as a function of $\overline{\chi} \triangleq E[E_S/N_0]$ for Nakagami-m fading with $m=1$.

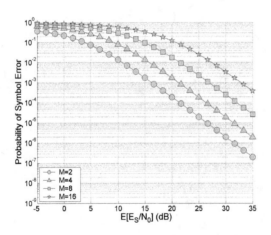

Fig. 103. Probability of symbol error $P_e = \overline{g}_M^D(\overline{\chi})$ as a function of $\overline{\chi} \triangleq E[E_S/N_0]$ for Nakagami-m fading with $m=2$.

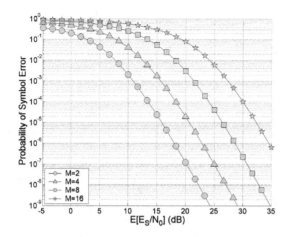

Fig. 104. Probability of symbol error $P_e = \overline{g}_M^D(\overline{\chi})$ as a function of $\overline{\chi} \triangleq E[E_S / N_0]$ for Nakagami-m fading with $m = 5$.

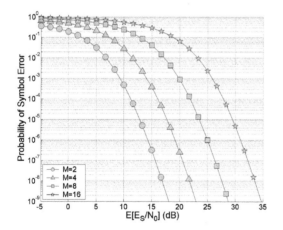

Fig. 105. Probability of symbol error $P_e = \overline{g}_M^D(\overline{\chi})$ as a function of $\overline{\chi} \triangleq E[E_S / N_0]$ for Nakagami-m fading with $m = 10$.

SNR estimation from the SER is done analogously to (182) and (183), namely, for case (i) we can estimate the instantaneous SNR via:

$$\eta_{dB}^D = 10 \cdot \log_{10}\left(\left(g_M^D\right)^{-1}(S(L))\right) \tag{209}$$

while for case (ii) we estimate the average SNR through:

$$\overline{\eta}_{dB}^D = 10 \cdot \log_{10}\left(\left(\overline{g}_M^D\right)^{-1}(S(L))\right) \tag{210}$$

For judging the efficacy of estimation via the SER, we ask: *What is the minimal value of L needed to ensure* $P\left(\left|\eta_{dB}^D - \chi_{dB}\right| < tol\right) > C$ (for case (i)) or $P\left(\left|\overline{\eta}_{dB}^D - \overline{\chi}_{dB}\right| < tol\right) > C$ (for case (ii)). From a derivation similar to that which led to (134), we find that for case (i) (i.e., $L \cdot T \ll T_{COH}$) the answer to this question is:

$$L > \frac{2 g_M^D\left(\chi\right)\left(1 - g_M^D\left(\chi\right)\right)\left(erf^{-1}(C)\right)^2}{\left(\min\left\{\left|g_M^D\left((1-w_2)\cdot\chi\right) - g_M^D\left(\chi\right)\right|, \left|g_M^D\left((1+w_1)\cdot\chi\right) - g_M^D\left(\chi\right)\right|\right\}\right)^2} \tag{211}$$

For case (ii) (i.e., $L \cdot T \gg T_{COH}$), the answer is:

$$L > \frac{2 \overline{g}_M^D\left(\overline{\chi}\right)\left(1 - \overline{g}_M^D\left(\overline{\chi}\right)\right)\left(erf^{-1}(C)\right)^2}{\left(\min\left\{\left|\overline{g}_M^D\left((1-w_2)\cdot\overline{\chi}\right) - \overline{g}_M^D\left(\overline{\chi}\right)\right|, \left|\overline{g}_M^D\left((1+w_1)\cdot\overline{\chi}\right) - \overline{g}_M^D\left(\overline{\chi}\right)\right|\right\}\right)^2} \tag{212}$$

where $w_1 \triangleq \left(10^{tol/10} - 1\right)$ and $w_2 \triangleq \left(1 - 10^{-tol/10}\right)$.

4.14.3 Graphical Exhibition of Results

Graphs of (204) vs. (211) are given in Fig. 106, where we see that the proposed estimator often requires considerably fewer symbol intervals in order to arrive at an equally accurate estimate. The lowest SNRs for which results are given in Fig. 106 are rough thresholds Γ_M defined as the SNRs at which $g_M^D(\Gamma_M) = 5 \cdot 10^{-2}$.

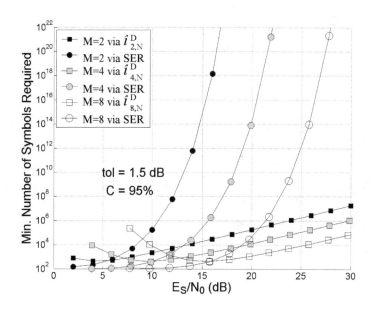

Fig. 106. Estimation via $\hat{l}^D_{M,N}$ vs. estimation via the SER for $2NT \ll T_{COH}$ and $L \cdot T \ll T_{COH}$.

In Fig. 107 to Fig. 110 we present results for case (b) and (ii), for Nakagami-m fading for various values of m. As can be seen in those figures, the advantage of the proposed technique is more pronounced for higher m's, and it is easily seen that as m increases the performance approaches that of case (a) vs. case (i), as shown in Fig. 106. This is explained by the fact that as $m \to \infty$ the Nakagami-m fading behaviour approximates a no-fading situation [13 Sec. 14.3]. Again, the lowest SNRs for which results are given in Fig. 107 to Fig. 110 are rough thresholds $\bar{\Gamma}_M$ defined as the average SNRs at which $\bar{g}^D_M(\bar{\Gamma}_M) = 5 \cdot 10^{-2}$.

It is important to make note of the fact that we presented here graphs of case (a) and case (i) (in Fig. 106) and case (b) and (ii) (in Fig. 107 to Fig. 110). Theoretically, there could be situations were L and N are such that one would have to compare case (a) to case (ii) or case (b) to case (i). Such a comparison can be made by looking at the appropriate curves taken from Fig. 106 to Fig. 110.

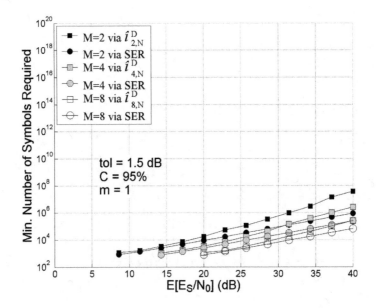

Fig. 107. Estimation via $\hat{i}_{M,N}^{D}$ vs. estimation via the SER, for $2NT \gg T_{COH}$ and $L \cdot T \gg T_{COH}$ for Nakagami-m fading with $m=1$.

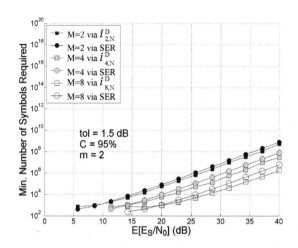

Fig. 108. Estimation via $\hat{l}_{M,N}^{D}$ vs. estimation via the SER, for $2NT \gg T_{COH}$ and $L \cdot T \gg T_{COH}$ for Nakagami-m fading with $m = 2$.

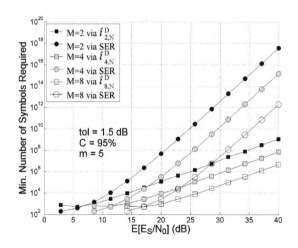

Fig. 109. Estimation via $\hat{l}_{M,N}^{D}$ vs. estimation via the SER, for $2NT \gg T_{COH}$ and $L \cdot T \gg T_{COH}$ for Nakagami-m fading with $m = 5$.

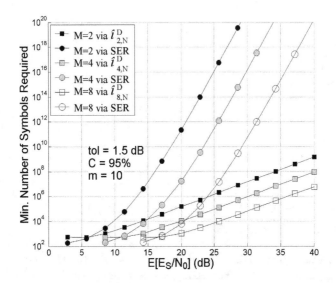

Fig. 110. Estimation via $\hat{l}_{M,N}^{D}$ vs. estimation via the SER, for $2NT \gg T_{COH}$ and $L \cdot T \gg T_{COH}$ for Nakagami-m fading with $m=10$.

4.14.4 Discussion of Results

The analysis of the results presented in Fig. 106 to Fig. 110 will proceed in a manner similar to that of Sec. 4.5.3. As can be seen by inspecting Fig. 106 to Fig. 110, the number of symbols needed for the estimation of the SNR from $\hat{l}_{M,N}^{D}$ does not change significantly as a function of the fading characteristics and coherence time. In contrast, estimation via the SER is strongly affected by the fading characteristics and the coherence time.

a) *Estimation latency analysis*

Let us first treat operation at high SNR. At high SNR we observe that estimation via $\hat{l}_{M,N}^{D}$ in general requires much less symbol intervals than estimation via the SER. This means that the proposed estimator can generate estimates much more rapidly than

- 214 -

estimation via the SER. An exception to this rule can be see in Fig. 107 (Nakagami-*m* fading with $m - 1$ (=Rayleigh fading)), where we see that estimation via the SER requires in general less symbols at high SNR. Inspection of Fig. 107 to Fig. 110 shows that as the fading index *m* increases, so does the advantage of the proposed estimator. For moderate and high m, and, as well, for case (a) ($2NT \ll T_{COH}$) we find that the proposed estimation method is much better than estimation via the SER, often by many orders of magnitude.

Now let us discuss low SNR operation. At low SNRs, we see that the proposed method requires about the same number of symbols as SNR estimation via the SER. Since it is often the case that the receiver spends most of its lifetime operating in the low-SNR region, one could make the argument that the advantage of the proposed method is minimal since it requires about the same number of symbol intervals as estimation via the SER. This, however, ignores several key issues. First, consider the case of unknown data being transmitted. The only way by which an SER estimate can be obtained from unknown data is by obtaining an error rate estimate from a code-decode process [62]. This means that one must first code the transmitted data at the transmitter and then decode it at the receiver (e.g., using block codes or convolution codes), and that in order to obtain an error rate estimate the receiver would compare the decoded data stream to the input data stream, hence arriving at an error rate estimate. This, however, implicitly assumes that the error correction decoder's output is completely error free – which is a fallacy at low SNRs. Hence, at lower SNRs the SER estimate would be inherently unreliable, with this problem being more severe as the SNR decreases. Moreover, the error correction decoder (ECD) may not even be locked at low SNRs, hence precluding SER estimation in the first place. To combat this problem, known symbols can be sent over the channel (in the form of training sequences, pilot symbols, or preambles) and the error rate estimation can be done upon those symbols. This, however, introduces two problems. First, obviously, the channel throughput that is taken up by those symbols cannot be used in order to transmit data, i.e. a reduction in the channel's information throughput is incurred. Secondly, unless we are prepared to significantly shut down the information-bearing content of the channel, the known symbols must only be allowed to take up a small percentage of the data stream. If we call this percentage *P* (e.g., *P*=10%), then we have that the number of symbol intervals that we actually have to wait in order to arrive at the SER estimate is increased by a factor of *1/P* over the quantities outlined in Fig. 106 to Fig. 110. For example, for $P=10\%$ we would need to multiply those quantities by a factor of 10, which clearly degrades the performance of estimation via the SER as compared to estimation via $\hat{l}_{M,N}^{D}$. Therefore, we can say that the results presented in Fig. 106 to Fig. 110 are optimistic with regards to estimation via the SER, and that, consequently, the proposed method is superior at low SNRs as well.

b) *Hardware complexity analysis*

In terms of complexity, we note that estimation via the SER requires the implementation of error detection and accrual mechanisms, which often necessitate a non-trivial amount of hardware and/or software resources. This, in addition to an algorithm or lookup table that would translate the SER measurement into an SNR measurement. In contrast, the proposed method is impervious to the content of the data stream, the coding method, and the error rate. Regarding fixed-point implementation of the proposed estimator, we make note of the fact that the value of *tol* must, obviously, be larger than the minimum resolution achievable given the quantization. In Sec. 4.14.3 we assumed that enough quantization bits were used and we ignore quantization effects (a good assumption, considering Sec. 4.12). Moreover, it is easily shown that accurate fixed-point hardware estimation of the SNR from the SER would require an unfeasibly large LUT (due to the large dynamic range of the SER). Thus, including quantization effects would have heavily favoured estimation via $\hat{l}_{M,N}^{D}$ even more.

Let us now delve even further into hardware complexity analysis. As noted in Section 4.11.1, in D-MPSK systems detection of the received symbols is often achieved via generating a pseudo-coherently demodulated M-PSK signal $u_n \triangleq r_n r_{n-1}^*$ (see [119 Sec. 6.5.3]). However, an equally valid detector would be via generation of the *normalized* pseudo-coherently demodulated M-PSK signal $v_n \triangleq \dfrac{r_n r_{n-1}^*}{|r_n||r_{n-1}|}$. In fact, it is trivial to see that the latter has advantages in terms of the stability of the dynamic range of the pseudo-coherent constellation vis-à-vis the AGC's operation. Thus, implementation of $v_n \triangleq \dfrac{r_n r_{n-1}^*}{|r_n||r_{n-1}|}$ in Fig. 96 obviates the need to generate the constellation $u_n \triangleq r_n r_{n-1}^*$, and, hence, it can be argues that the only real hardware penalty incurred by implementing the

- 216 -

proposed estimator is the sequence LUT#3-IAD-LUT#4 (see Fig. 96), which is the same order of complexity as the estimator of Part A, i.e., trivial.

4.15   Comparison to other estimators using the NMSE metric

In Sec. 4.13-4.14 we focused on comparisons of the proposed method versus estimation via the SER. As mentioned in Section 4.14, this is due to the fact that estimation via the SER is a prevalent and universally applicable SNR estimation method. Other SNR estimators have been suggested in [59], [60], [61], [62], [63], [64], [65], [66], [67], [68], [69], [70], [71], [72], [73], [16] and [74]. In theory, these too could be applied to the pseudo-coherent constellation v_n in order to produce an SNR estimate, although such a procedure is not immediate given that the noise statistics of v_n differ from those of r_n. While in-depth comparison vs. all those estimators is impossible within the span of the current book, it is claimed that the proposed estimator possesses several qualitative advantages which would indicate a favourable outcome to such a comparison. Those qualitative advantages were highlighted in Sec. 4.6.1 and are equally applicable in the current case.

4.15.1   Comparison for $2NT \ll T_{COH}$

For completeness, we now present NMSE and normalized bias results vs. the M_2M_4 and SVR estimators. The comparison vs. the M_2M_4 estimator is particularly important because, as outlined in Sec. 4.6, the M_2M_4 estimator is a blind estimator operating on 1 sample/symbol, and is perhaps the best previously available estimator of this kind [60]. As noted in Sec. 4.6, the M_2M_4 and SVR estimators are impervious to the carrier phase, and therefore their performance results for D-MPSK are identical to those for M-PSK. In contrast, estimation from $\hat{l}_{M,N}^{D}$ will have different performance than that of $\hat{l}_{M,N}$, as we have already seen in Sec. 4.13-4.14. We shall now present NMSE and normalized bias comparisons vs. the M_2M_4 estimator and SVR estimators.

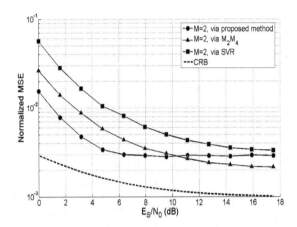

Fig. 111. NMSE comparison of estimation via $\hat{l}^D_{M,N}$ vs. the M₂M₄ estimator and the SVR estimator, $M=2$, with $2N=1024$ symbols used to compute each estimator.

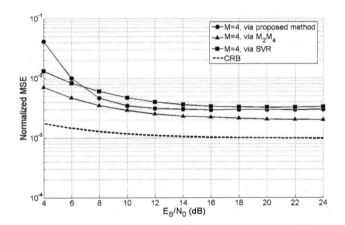

Fig. 112. NMSE comparison of estimation via $\hat{l}^D_{M,N}$ vs. the M₂M₄ estimator and the SVR estimator, $M=4$, with $2N=1024$ symbols used to compute each estimator.

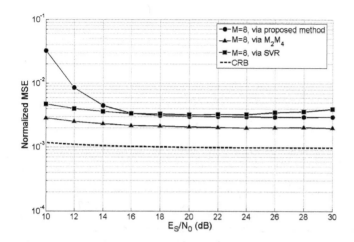

Fig. 113. NMSE comparison of estimation via $\hat{l}_{M,N}^{D}$ vs. the M₂M₄ estimator and the SVR estimator, $M=8$, with $2N=1024$ symbols used to compute each estimator.

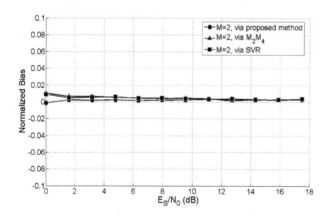

Fig. 114. Normalized bias comparison of estimation via $\hat{l}_{M,N}^{D}$ vs. the M₂M₄ estimator and the SVR estimator, $M=2$, with $2N=1024$ symbols used to compute each estimator.

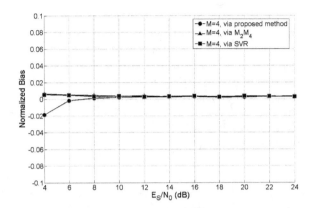

Fig. 115. Normalized bias comparison of estimation via $\hat{l}_{M,N}^{D}$ vs. the M₂M₄ estimator and the SVR estimator, $M=4$, with $2N=1024$ symbols used to compute each estimator.

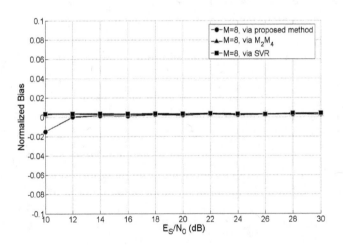

Fig. 116. Normalized bias comparison of estimation via $\hat{l}_{M,N}^{D}$ vs. the M₂M₄ estimator and the SVR estimator, $M=8$, with $2N=1024$ symbols used to compute each estimator.

First, let us discuss the NMSE results. As can be seen in Fig. 111-Fig. 113 estimation via $\hat{l}^D_{M,N}$ performs very respectfully at medium and high SNR. At such SNRs, it is better than the SVR estimator and only slightly worse than the M_2M_4 estimator. We can see that at high SNR estimation via $\hat{l}^D_{M,N}$ tends to a high-SNR bound that is 50% higher than estimation via the M_2M_4 estimator. This is not surprising, since $\hat{l}^D_{M,N}$ operates on the pseudo-demodulated constellation, whose phase perturbation variance is higher than that of the original constellation upon which the M_2M_4 estimator operates (compare (192) to (33)). However, this disadvantage may be overcome if the N used to compute $\hat{l}^D_{M,N}$ is simply increased by 50%. Regarding operation at low SNRs, the NMSE may be decreased by increasing N, with the tradeoff being a longer estimation period. Such a tradeoff may be acceptable, since (as noted in Sec. 4.8 in comments that are easily applied here) an increase in N causes a negligible increase in the complexity of $\hat{l}^D_{M,N}$. Thus, if the designer of the system for which SNR estimates are produced is more concerned about hardware efficiency than about estimation latency, estimation via $\hat{l}^D_{M,N}$ for D-MPSK (or M-PSK in the case of lack of carrier synchronization) may be an attractive choice over the M_2M_4 estimator (for coherent M-PSK when the carrier PLL is locked, estimation via $\hat{l}_{M,N}$ is a much better alternative to both, as seen in Sec. 4.6).

4.15.2   Comparison for $2NT \gg T_{COH}$

When $2NT \gg T_{COH}$, then estimation is of the average SNR and that estimate is obtained through $\overline{\gamma}^D_{dB} = 10 \cdot \log_{10}\left(\left(\overline{\hat{J}}^D_M\right)^{-1}(\hat{l}^D_{M,N})\right)$ (see Sec. 4.13). As noted in Sec. 4.8, the M_2M_4 estimator performs very poorly under fading conditions [70], and in general more complicated method based on the Viterbi algorithm [15] or EM algorithm [70] have been suggested for SNR estimation in fading conditions. In order to nonetheless evaluate the effects of fading upon the proposed estimator's NMSE, simulation results are obtained for the case of Nakagami-m fading for various values of m, and the NMSE is compared to the NMSE without fading.

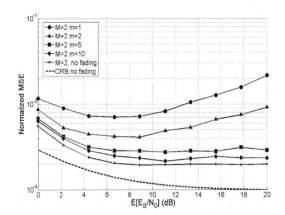

Fig. 117. NMSE comparison of SNR estimation via estimation via $\hat{l}_{M,N}^{D}$ for $M=2$ and Nakagami-m fading. For each metric $2N=1024$ symbols were used.

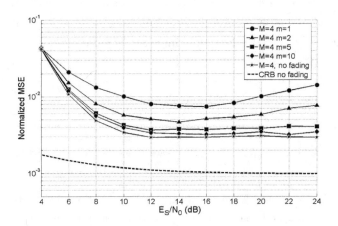

Fig. 118. NMSE comparison of SNR estimation via estimation via $\hat{l}_{M,N}^{D}$ for $M=4$ and Nakagami-m fading. For each metric $2N=1024$ symbols were used.

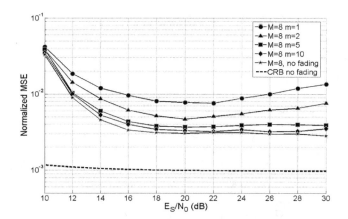

Fig. 119. NMSE comparison of SNR estimation via estimation via $\hat{l}_{M,N}^D$ for $M = 8$ and Nakagami-m fading. For each metric $2N = 1024$ symbols were used.

We see from Fig. 117-Fig. 119 that over much of the SNR range of interest an increase in the NMSE is observed at (particularly at high-SNR, although this is not particularly problematic since SNR estimates usually need not be very accurate at high SNRs, since only minor performance gains of coders/equalizers are achievable at high SNRs from precise knowledge of the SNR). As noted in Sec. 4.8, the NMSE can be reduced simply by increasing the number of symbols $2N$ that is used in order to compute the estimate, which will have a negligible effect on the estimator's complexity.

We now compare the biases of estimation via $\hat{l}_{M,N}^D$ under fading conditions to that which is obtained when fading is not present or when $2NT \ll T_{COH}$.

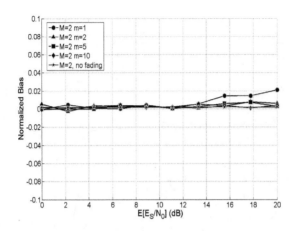

Fig. 120. Normalized bias comparisons for SNR estimation via $\hat{l}_{M,N}^D$ for $M=2$ and Nakagami-m fading. For each metric $2N=1024$ symbols were used.

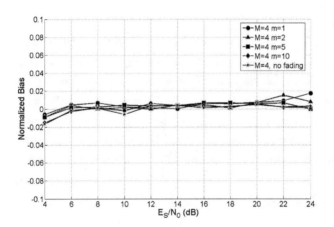

Fig. 121. Normalized bias comparisons for SNR estimation via $\hat{l}_{M,N}^D$ for $M=4$ and Nakagami-m fading. For each metric $2N=1024$ symbols were used.

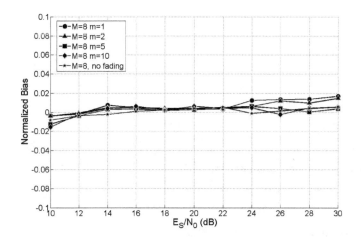

Fig. 122. Normalized bias comparisons for SNR estimation via $\hat{l}^D_{M,N}$ for $M=8$ and Nakagami-m fading. For each metric $2N=1024$ symbols were used.

As can be seen in Fig. 120-Fig. 122, the bias of the proposed estimator is very near the optimal value of 0 for all fadings and SNRs considered.

4.16 Application to SNR Estimation for M-PSK in the Absence of Carrier Synchronization

A very important observation is that the proposed SNR estimator can be used to provide an SNR estimate for *coherent* M-PSK but *without the need for carrier synchronization* (which is a prerequisite for the estimator of Part A). We note, however, that in general SNR estimation for coherent M-PSK via $\hat{l}^D_{M,N}$ requires more symbol intervals than estimation via $\hat{l}_{M,N}$ (see Fig. 69 to Fig. 71, Fig. 85 to Fig. 88, Fig. 106 to Fig. 110). Thus, the estimator of Part A should be used when carrier synchronization has been achieved. It is emphasized that for SNR estimation via $\hat{l}^D_{M,N}$ to be done on an M-PSK signal in the absence of carrier synchronization, there is no need to differentially code or decode the datastream (since the estimator is NDA). Finally, we note that for

- 225 -

estimation via $\hat{l}^D_{M,N}$ to work, the same limitation on the carrier error exists as was the case for the D-MPSK, namely that $|\Delta\omega| \ll 2\pi/(M \cdot T)$. We note that this is not a very problematic limitation since for the carrier PLL to lock the frequency error must be within the lock range of the PLL which is $\pm\Delta\omega_L = \pm 2\zeta\omega_n$ (see [103 Chap. 2]) and since in typical PLLs $\omega_n \ll 1/T$ and $0.7 \le \zeta \le 1.3$ (see Sec. 2.4 and [25 Chap. 7]) we then typically have that $|\Delta\omega| \le 2\zeta\omega_n \ll 2\pi/(M \cdot T)$ when the carrier PLL is near or within its lock range. However, when the carrier PLL is in search mode, i.e. the VCO or NCO is made to scan the potentially much wider frequency uncertainty, then the condition $|\Delta\omega| \ll 2\pi/(M \cdot T)$ might not hold, and care must be exercised if SNR estimation from $\hat{l}^D_{M,N}$ is attempted.

4.17 Conclusions

In this chapter, we started off in Part A by presenting an analysis of a new method of E_S/N_0 estimation for M-PSK receivers, and quantitative formulas describing its performance were developed. It was found that the proposed method has several quantitative and qualitative advantages with respect to previously available methods. These include: independence from the received data, no reliance on symbol decisions or error detection, resilience to AGC imperfections, a simple and compact fixed-point hardware implementation, and, perhaps most importantly, it requires only a relatively small number of symbols to arrive at an accurate estimate (as compared to SNR estimation via the SER). This method is therefore particularly well-suited for real-time estimate generation in M-PSK receivers. We presented quantitative NSME comparisons vs. the M_2M_4 and SVR estimators and found that the proposed estimator has respectable and often better performance in comparison. We also provided quantitative results for the performance of the proposed estimator in Nakagami-m fading.

In the second part of this chapter we presented a new Non Data Aided SNR estimator for D-MPSK operating at 1 sample/symbol. It was found that the estimator has a simple fixed-point hardware implementation that can be easily implemented within contemporary FPGAs or ASICs. Quantitative comparisons were conducted vs. the M_2M_4 and SVR estimators as well as estimation from the SER. These results included investigation of the estimator's performance in Nakagami-m fading. In terms of speed

- 226 -

and accuracy performance of the estimator, it was shown that it generally performs much better that estimation via the SER and also possesses several qualitative and quantitative implementational advantages. This estimator was also compared in the NMSE sense to the M_2M_4 and SVR estimators, and was found to be a competitive estimator (though less so than the estimator discussed in Part A). For all of the preceding reasons, the SNR estimation method proposed in Part B in this chapter has immediate applications in contemporary D-MPSK communications systems as well as in SNR estimation for M-PSK systems when carrier synchronization has not yet been achieved.

In a coherent M-PSK receiver, the SNR estimator of Part A of this chapter is an attractive choice once carrier synchronization has been achieved. As shown, the estimator of Part A uses less symbol intervals as compared to the estimator of Part B and thus should be preferred over the latter when the carrier PLL is locked. However, if the carrier PLL is unlocked, it was shown that the signal's SNR can be estimated using the estimator of Part B. Hence, to obtain an SNR estimate in coherent M-PSK receivers during both the tracking and acquisition operation modes of the carrier PLL, the M-PSK receiver would use the estimator of Part A when the carrier PLL is locked, and opt to use the estimator of Part B when the carrier is unlocked.

Chapter 5   Conclusions and Future
Work

5.1   Summary of Contributions

In this book, we presented new structures for lock detection, phase detection, and SNR estimation in M-PSK receivers. In Chapter 2 we presented a new type of self-normalizing lock detectors for carrier synchronization PLLs in M-PSK receivers. This was followed in Chapter 3 by presentation of two new families of phase detectors, and in Chapter 4 two new SNR estimation structures were defined and analyzed.

A recurring theme in all of the structures is that they are self-normalizing. Sometimes, when the normalization term is omitted the proposed structure reduces to previous known structures. For example, the normalized M^{th}-order nonlinearity presented in Sec. 3.4,   namely $d_M(n) \triangleq \mathrm{Im}[(I(n)+jQ(n))^M] \Big/ \left(I^2(n)+Q^2(n)\right)^{\frac{M}{2}}$, is reduced to the non-normalized M^{th}-order nonlinearity phase detector $c_M(n) \triangleq \mathrm{Im}[(I(n)+jQ(n))^M]$ when the denominator term is omitted. Similar observations can be done with regards to the lock detector presented in Chapter 2, which reduces to the Mth-order nonlinearity lock detector when the normalization term is omitted.

Despite their relative simplicity, it was shown that the normalization factors have a profound effect upon the proposed structures' behaviour and implementation. It was shown that (in contrast to non-normalized structures) the proposed structures are very resilient to vis-à-vis the AGC's operating point and performance. Furthermore, the normalization reduced the dynamic range of the proposed structures, which allows them to be efficiently implemented in fixed-point hardware, which is an issue that was given specific attention in this book. These efficient implementations contrast with the considerably more complicated implementations of previously available structures, a subject that also was discussed in this book.

As an additional advantage, it was found that the proposed structures are interrelated and often these interrelationships can aid in the theoretical analysis and can be exploited in order to achieve better performance. For example, the lock detector in Chapter 2 was shown to be a good estimator of the gain of the phase detector of Sec. 3.4, and this was exploited in Sec. 3.6 in order to construct a constant-gain phase detector. As another example, the SNR estimator of Part A in Chapter 4 is based upon the lock detector of Chapter 2, and in turn the SNR estimator of Part B of Chapter 4 is based upon an enhancement of the SNR estimator of Part A of Chapter 4. These interrelationships were exploited throughout Chapter 4 in order to arrive at qualitative and quantitative results.

It can be said that there are two uniting thread of this book. The first being that all the structures presented have a compact and practical implementation. Indeed, a book in an engineering discipline has only limited value if it cannot be applied in the real world. Hence, special emphasis was given in order to find and analyze practical implementations for the proposed structures. The second unifying thread of this book has been the effort to achieve results that will be useful to the practicing engineer on an intuitive level. This has been the motivation, for example, in finding and analyzing the closed-form approximate value of the expectation of the lock detector of Chapter 2, namely $f_M(\chi) \approx \exp\left(-M^2/4\chi\right)$ (eq. (35)), which can be computed by hand by the engineer and nonetheless serves as a very accurate predictor of the lock-detector value. A similar desire for practical usefulness motivated the derivation and analysis of the approximate results given in Chapter 3 and Chapter 4, and the inclusion of sections that treat the subject matter on an intuitive level, such as Sec. 1.5, 2.3.6, and 3.4.10. Moreover, the choice of quantitative comparison metrics in Chapter 4 (see Sec. 4.5, Sec. 4.14) was specifically motivated by the usefulness of such comparisons in a practical engineering situation, as explained in Sec. 4.4. Although this book includes exact results for all of the structures analyzed, the approximate but simple results are arguably more important in an engineering environment, since good but conceptually complicated structures are often rejected at the design stage because engineers often prefer to use inferior but tractable structures. Hence, it is hoped that the inclusion of approximate but simple expressions in this book will facilitate the proposed structures' adoption by the grassroots engineering community so that they will not remain solely in the realm of academia.

5.2 Future Work

5.2.1 Future Research: Analysis during Lack of Symbol Synchronization

There is significant and potentially fruitful terrain for future research based upon this book. One obvious avenue of such research is the investigation of the proposed structures when the symbol synchronization PLL is unlocked. Indeed, it can be shown through simulations that the proposed structures will indeed work when the symbol synchronization PLL is unlocked, but that the lack of symbol synchronization will cause significant changes in the structures' statistical properties. A complicating factor is that it is easily seen, even intuitively, that when the receiver lacks symbol synchronization the value of the proposed structures is highly dependent upon the symbols' baseband pulse shape. This contrasts with the situation discussed in this book, in which it was shown that the performance of the proposed structures when the symbol PLL is locked is independent of the baseband pulse shape so long as the post-matched-filter pulse shape conforms to the Nyquist criterion for zero ISI [13 Sec. 9.2.1]. Nonetheless, the path to analyzing the performance of the proposed structures in the absences of symbol synchronization is relatively straightforward, at least conceptually.

5.2.2 Future Research and Continuing Research: A Constant-Gain Detector during Both Tracking and Acquisition

An almost obvious extension of the work in this book is the construction of a phase detector that has constant gain during both tracking and acquisition. The principle underlying this detector is simple. In Sec. 3.6 we estimated the gain of $d_M(n)$ using $M \cdot \hat{I}_{M,N}$ and achieved a constant-gain detector during tracking via $d_M(n)/(M \cdot \hat{I}_{M,N})$. To achieve a constant-gain detector during acquisition, i.e. when the carrier is unlocked, we can employ a similar procedure by estimating[20] the gain of $d_M(n)$ using $M \cdot \left(f_M \circ \left(f_M^D \right)^{-1} \right) \left(\hat{l}_{M,N}^D \right)$, and then arriving at a constant-gain detector during acquisition via $d_M(n) \Big/ \left(M \cdot \left(f_M \circ \left(f_M^D \right)^{-1} \right) \left(\hat{l}_{M,N}^D \right) \right)$. This has the potential of allowing the carrier PLL to perform optimally not only during tracking, but also during acquisition. This is investigated in depth in [104].

5.2.3   Future and Continuing Research: Symbol Synchronization PLL Structures

Another proven avenue of research based upon the current book is the investigation of similar self-normalizing structures for the symbol timing synchronization PLL.

a)   *Revised system model accounting for lack of symbol timing synchronization*

In order to give an outline of this field, we first modify the system model given in Sec. 1.4 in order to account for the possible lack of symbol synchronization . The revised model is as follows.

The baseband M-PSK signal before modulation is defined as $m(t) \triangleq \sum_{r=-\infty}^{\infty} \exp\left(j\phi_r \right) p(t - rT)$ where $1/T$ is the symbol rate, $\phi_r \triangleq 2\pi \cdot m_r / M + \Pi_M \cdot \pi / M$, $m_r \in \{0, 1, ..., M-1\}$, and $\Pi_M \triangleq \{1$ if $M \neq 2$, 0 if $M = 2\}$. We use the notation τ_i to signify the signal's propagation delay. At the input of the I-Q demodulator the IF signal is $s_m(t) = \mathrm{Re}[m(t - \tau_i)\exp(j\omega_i t + j\theta_i)]$ and that signal is corrupted by AWGN. The revised M-PSK model which includes the possible effects of the lack of symbol synchronization is as shown in Fig. 123, where:

1. $1/T_s = 2/T$ is the sample rate.

---

[20]   In the general discussion in this section we assume that no fading is present. However, in [104] fading effects are treated.

2. $n(t) \sim N(0, N_0 W)$ where W is the width of the bandpass IF filter before the I-Q demodulator (not shown).

3. We assume a narrowband bandpass signal (i.e. $\omega_i \gg 1/T$) and that the Nyquist criterion for zero-ISI [13 Sec. 9.2.1] is obeyed regarding the output of the matched filters.

4. K_I and K_Q are the equivalent gains associated with the circuit, and are a slow function of time controlled by the AGC circuit (the AGC's purpose is to ensure that the dynamic range of the samplers is utilized yet the samplers are not saturated). For simplicity in this book we assumed that these gains are equal, i.e. $K_I = K_Q = K$ (which is usually the case), but this is not a necessary requirement for the symbol-PLL structures that we shall shortly outline.

5. The matched filter $h(t) = p(-t)$ is assumed ideal.

6. When the carrier loop is locked we have $\Delta\omega = 0$ and (since M-PSK carrier synchronization has an inherent M-fold phase ambiguity ([22 Chap. 5, 6], [21 Sec. 5.7])) $\theta_o \in \{\theta_i + 2\pi k / M - \theta_e | k = 0,1,...,M-1\}$, where $|\theta_e| < \pi / M$ is the residual carrier phase error.

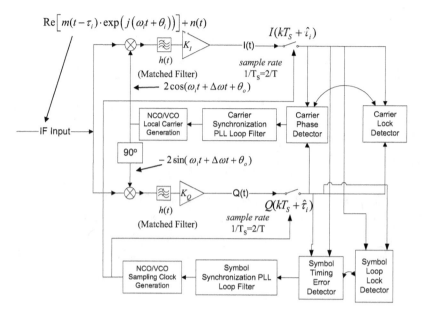

Fig. 123. General structure of a coherent M-PSK receiver showing both the carrier and symbol PLLs.

The notation $\hat{\tau}_i$ is employed to refer to the receiver's estimate of τ_i. The symbol synchronization timing error is defined as $\tau \triangleq (\tau_i - \hat{\tau}_i) \bmod_T$, with $\tau \in [-\frac{1}{2}, \frac{1}{2}]$. The even samples of the channels are then:

$$I_e(n) \triangleq I(t)\big|_{t=2nT_s + \hat{\tau}_i} \quad \text{and} \quad Q_e(n) \triangleq Q(t)\big|_{t=2nT_s + \hat{\tau}_i} \tag{213}$$

and the odd samples are:

$$I_o(n) \triangleq I(t)\big|_{t=(2n+1)T_s + \hat{\tau}_i} \quad \text{and} \quad Q_o(n) \triangleq Q(t)\big|_{t=(2n+1)T_s + \hat{\tau}_i} \tag{214}$$

It is worth noting that under perfect symbol synchronization conditions (that is, $\hat{\tau}_i = \tau_i$), the even samples correspond to the peaks of the symbols, and the odd samples correspond to the transitions between symbols.

- 233 -

b) Symbol PLL lock detection and SNR estimation

We can apply the principle of self-normalization used in this book in order to arrive at a new lock detector for the symbol PLL that is based upon the non-normalized detector of Karam et al. [120], which is (detector B in [120]) $\frac{1}{2N}\sum_{n=-N+1}^{N}\left(\left(I_e^2(n)-I_o^2(n)\right)+\Pi_M\cdot\left(Q_e^2(n)-Q_o^2(n)\right)\right)$. We define the normalized symbol PLL lock detector as:

$$s_N \triangleq \frac{1}{2N}\sum_{n=-N+1}^{N}\left(\left[\frac{I_e^2(n)-I_o^2(n)}{I_e^2(n)+I_o^2(n)}\right]+\Pi_M\left[\frac{Q_e^2(n)-Q_o^2(n)}{Q_e^2(n)+Q_o^2(n)}\right]\right) \tag{215}$$

Since the lock detector presented in is a one-to-one function of the SNR, the SNR can be estimated from the lock detector value, just like an SNR estimate was generated from the carrier PLL lock metric value in Chapter 4 of this book. This is explored in [121]. Unlike the estimators of Chapter 4, which were independent of the post-matched filter pulse shape so long as the latter conformed to the Nyquist criterion for zero ISI, estimators based upon the symbol PLL lock detector are dependent upon the pulse shape. Define

$$f_{M,p}(\chi) \triangleq E[s_N \mid E_s / N_0 = \chi, \text{Tx pulse shape is p(t)}] \tag{216}$$

then the SNR can be estimated via:

$$\gamma_{dB}^S = 10\cdot\log_{10}\left(f_{M,p}^{-1}(s_N)\right) \tag{217}$$

The definition and investigation of this symbol-PLL lock detector and SNR estimator structure was done in [122], [123], and [121] (in [122] and [123] only BPSK and QPSK are considered, though extension to operation for $M > 4$ is straightforward and is discussed in [121]). An efficient fixed-point hardware implementation of s_N is shown in Fig. 124. Obviously, for actual symbol PLL lock detection to occur, s_N must be compared ([122],[123]) to a lock threshold (not shown in Fig. 124).

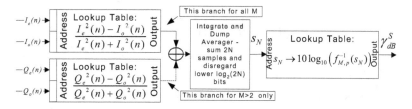

Fig. 124. Efficient fixed-point hardware implementation of the symbol synchronization PLL lock detector s_N ([122], [123], [121]) and associated SNR estimation method [121].

c) Timing error detectors for the symbol PLL

Using the principle of self-normalization, we can derive new Timing Error Detectors (TEDs) for the symbol PLL (a illustration of where the TED operates in the M-PSK receiver, see Fig. 123). For example, we can start off from the Gardner TED [42], defined for QPSK as:

$$g_{QPSK}(n) \triangleq \left(I_e(n) - I_e(n-1)\right)I_o(n-1)$$
$$+\left(Q_e(n) - Q_e(n-1)\right)Q_o(n-1)$$

(218)

Using normalization, we arrive at a new detector:

$$\tilde{g}_{QPSK}(n) \triangleq \frac{I_e(n)I_o(n-1)}{\left(I_e^2(n) + I_o^2(n-1)\right)} - \frac{I_e(n-1)I_o(n-1)}{\left(I_e^2(n-1) + I_o^2(n-1)\right)}$$
$$+ \frac{Q_e(n)Q_o(n-1)}{\left(Q_e^2(n) + Q_o^2(n-1)\right)} - \frac{Q_e(n-1)Q_o(n-1)}{\left(Q_e^2(n-1) + Q_o^2(n-1)\right)}$$

(219)

This detector is investigated in [43], and it is shown that it possesses significant implementational and performance advantages vis-à-vis (218).

Another detector can be derived from the Mueller & Müller DD (M&M) [34] (1 sample/symbol), for example for BPSK:

$$MM(n) = sign(I_e(n-1))I_e(n) - sign(I_e(n))I_e(n-1)$$

(220)

and the corresponding normalized detector is:

$$\widetilde{MM}(n) \triangleq \frac{sign(I(n-1))I(n) - sign(I(n))I(n-1)}{\sqrt{I^2(n-1) + I^2(n)}} \qquad (221)$$

This detector, as well as a normalized version of the Decision Directed Gardner detector, was analyzed in [37], which also includes definition, analysis and results for the new normalized detectors for $M>2$. Once again, as is shown in [37], the normalized detectors have many performance and implementation advantages.

5.3   Final Remarks

As was shown in this concluding chapter, the structures proposed in this book have immediate practical applications in current M-PSK receivers. Moreover, the principles used to derive these new structures can be applied to the symbol timing synchronization PLLs, where similar performance and implementation advantages are observed. A promising avenue of future work is the investigation of the structures proposed in this book when the symbol synchronization PLL is unlocked. In short, the structures and methods proposed in this book have immediate applications as well as significant promise of future discovery.

References

[1]   Y. Linn, "A Methodical Approach to Hybrid PLL Design for High-Speed Wireless Communications," in *Proc. 8th IEEE Wireless and Microwave Technology Conf. (WAMICON 2006)*, Clearwater, FL, Dec. 4-5, 2006.

[2]   K. Young-Wan, C. Jong-Suk, and P. Dong-Chul, "Circuit design and performance analysis of carrier recovery loop for digital DBS system in the presence of phase noise," *IEEE Trans. Broadcasting*, vol. 45, no. 3, pp. 294-302, Sep. 1999.

[3]   ETSI (European Telecommunications Standards Institute), "DVB-S2 Technical Report ETSI TR 102 376 V1.1.1," Feb. 2005.

[4]   ETSI (European Telecommunications Standards Institute), "Implementation guidelines for DVB terrestrial services: Technical Report ETSI TR 101 190 V1.2.1 ", Nov. 2004.

[5]   ETSI (European Telecommunications Standards Institute), "DVB-H Implementation Guidelines: Technical Report ETSI TR 102 377 V1.2.1," Nov. 2005.

[6]   B. O'Hara and A. Petrick, *The IEEE 802.11 handbook : a designer's companion*, 2nd ed. New York: IEEE, 2004.

[7]   J. P. K. Gilb, *Wireless multimedia : a guide to the IEEE 802.15.3 standard.* New York, NY: Standards Information Network, IEEE Press, 2004.

[8]   IEEE 802.16 Working Group, "IEEE Standard for Local and Metropolitan Area Networks Part 16: Air Interface for Fixed Broadband Wireless Access Systems (IEEE Std 802.16-2004)," Oct. 2004.

[9]   W. P. Osborne and B. T. Kopp, "An analysis of carrier phase jitter in an M-PSK receiver utilizing MAP estimation," in *Proc. MILCOM '93*, Boston, MA, USA, 1993, pp. 465-470.

[10]   M. M. J. L. van de Kamp, "Climatic radiowave propagation models for the design of satellite communication systems," PhD. Thesis, Technische Universiteit Eindhoven, The Netherlands, 1999.

[11]   R. M. Gagliardi, *Satellite communications. Second edition*: Van Nostrand Reinhold, 1991.

[12]   T. A. Summers and S. G. Wilson, "SNR mismatch and online estimation in turbo decoding," *IEEE Trans. Commun.*, vol. 46, no. 4, pp. 421-423, Apr. 1998.

[13]   J. G. Proakis, *Digital communications*, 4th ed. Boston: McGraw-Hill, 2001.

[14]   ETSI (European Telecommunications Standards Institute), "DVB-S2 Technical Report ETSI TR 102 376 V1.1.1," 2005.

[15]   K. Balachandran, S. R. Kadaba, and S. Nanda, "Channel quality estimation and rate adaptation for cellular mobile radio," *IEEE Journal on Selected Areas in Communications*, vol. 17, no. 7, pp. 1244-1256, Jul. 1999.

[16]   D. R. Pauluzzi, "Signal-to-noise ratio and signal-to-impairment ratio estimation in AWGN and wireless channels," Ph.D. Thesis, Queen's University, Kingston, Ontario, Canada, Sept. 1997.

[17]   R. G. Lyons, *Understanding digital signal processing*, 2nd ed. NJ: Prentice Hall, 2004.

[18]   Y. Linn, "A Tutorial on Hybrid PLL Design for Synchronization in Wireless Receivers," in *Proc. International Seminar: 15 Years of Electronic Engineering*, Universidad Pontificia Bolivariana, Bucaramanga, Colombia, Aug. 15-19, 2006 *(invited paper)*.

[19]   M. K. Simon, S. M. Hinedi, and W. C. Lindsey, *Digital communication techniques*. NJ: Prentice Hall, 1995.

[20]   M. K. Simon and D. Divsalar, "Some new twists to problems involving the Gaussian probability integral," *IEEE Trans. Commun.*, vol. 46, no. 2, pp. 200-210, Feb. 1998.

[21]   U. Mengali and A. N. D'Andrea, *Synchronization techniques for digital receivers*. NY: Plenum Press, 1997.

[22]   H. Meyr, M. Moeneclaey, and S. Fechtel, *Digital communication receivers: synchronization, channel estimation, and signal processing*. NY: Wiley, 1998.

[23]   P. K. Vitthaladevuni and M. S. Alouini, "Effect of imperfect phase and timing synchronization on the bit-error rate performance of PSK modulations," *IEEE Trans. Commun.*, vol. 53, no. 7, pp. 1096-1099, Jul. 2005.

[24]   W. Lindsey and M. Simon, "The Effect of Loop Stress on the Performance of Phase-Coherent Communication Systems," *IEEE Trans. Commun.*, vol. 18, no. 5, pp. 569-588, Oct. 1970.

[25]   F. M. Gardner, *Phaselock techniques*, 2nd ed. NY: Wiley, 1979.

[26]   W. C. Lindsey and M. K. Simon, *Telecommunication systems engineering*. NJ: Prentice-Hall, 1973.

[27]   S. Haykin, *Communication systems*, 2nd ed. NY: Wiley, 1983.

[28]   D. R. Stephens, *Phase-locked loops for wireless communications : digital, analog, and optical implementations*, 2nd ed. Boston: Kluwer Academic, 2002.

[29]   D. H. Wolaver, *Phase-locked loop circuit design*. NJ: Prentice Hall, 1991.

[30]   E. C. Ifeachor and B. W. Jervis, *Digital signal processing: a practical approach*, 2nd ed. NY: Prentice Hall, 2002.

[31]   A. V. Oppenheim and R. W. Schafer, *Discrete-time signal processing*. NJ: Prentice Hall, 1989.

[32]   F. M. Gardner, "Interpolation in digital modems. I. Fundamentals," *IEEE Trans. Commun.*, vol. 41, no. 3, pp. 501-507, Mar. 1993.

[33]   L. Erup, F. M. Gardner, and R. A. Harris, "Interpolation in digital modems. II. Implementation and performance," *IEEE Trans. Commun.*, vol. 41, no. 6, pp. 998-1008, Jun. 1993.

[34]   K. H. Mueller and M. Muller, "Timing recovery in digital synchronous data receivers," *IEEE Trans. Commun.*, vol. 24, no. 5, pp. 516-530, May 1976.

[35]   W. G. Cowley and L. P. Sabel, "The performance of two symbol timing recovery algorithms for PSK demodulators," *IEEE Trans. Commun.*, vol. 42, no. 6, pp. 2345-2355, Jun. 1994.

[36]   M. Moeneclaey and T. Batsele, "Carrier-independent NDA symbol synchronization for M-PSK, operating at only one sample per symbol," in *Proc. GLOBECOM '90*, San Diego, CA, USA, 1990, pp. 594-598.

[37]   Y. Linn, "Two new decision directed M-PSK timing error detectors," in *Proc. 18th Canadian Conference on Electrical and Computer Engineering (CCECE'05)*, Saskatoon, SK, Canada, May 1-4, 2005, pp. 1759-1766.

[38]   D. Verdin and T. C. Tozer, "Symbol-timing recovery for M-PSK modulation schemes using the signum function," in *Proc. IEE Colloquium on New Synchronisation Techniques for Radio Systems (Digest No.1995/220)*, London, UK, 1995, pp. 2/1-2/7.

[39]   D. Verdin, "Synchronization in sampled receivers for narrowband digital modulation schemes," Ph.D. Thesis, in *Dept. of Electrical Engineering*, University of York, United Kingdom, 1996.

[40]   L. P. Sabel, "A Maximum Likelihood Approach to Symbol Timing Recovery in Digital Communications," Ph.D. Thesis, in *School of Electronic Engineering*, University of South Australia, Adelaide, Australia, 1993.

[41]   Y. Linn, "Synchronization and Receiver Structures in Digital Wireless Communications (workshop notes)," in *International Seminar: 15 Years of Electronic Engineering*. Universidad Pontificia Bolivariana, Bucaramanga, Colombia, Aug. 15-19, 2006.

[42]   F. M. Gardner, "A BPSK/QPSK timing-error detector for sampled receivers," *IEEE Trans. Commun.*, vol. 34, no. 5, pp. 423-429, May 1986.

[43]   Y. Linn, "A new NDA timing error detector for BPSK and QPSK with an efficient hardware implementation for ASIC-based and FPGA-based wireless receivers," in *Proc. 2004 IEEE Intl. Symp. on Circuits and Systems (ISCAS'04)*, Vancouver, BC, Canada, May 23-26, 2004, pp. IV:465-468.

[44]   D. Verdin and T. C. Tozer, "Symbol timing recovery scheme tolerant to carrier phase error," *Electronics Letters*, vol. 30, no. 2, pp. 116-117, Jan. 1994.

[45]   R. L. Peterson, R. E. Ziemer, and D. E. Borth, *Introduction to spread-spectrum communications*. NJ: Prentice Hall, 1995.

[46]   Analog Devices, "AD9851 Datasheet, Rev. C," retrieved from www.analog.com.

[47]   J. D. Gibson (editor), "The communications handbook," 2nd ed. Boca Raton, FL: CRC Press, 2002.

[48]   A. Blanchard, *Phase-locked loops. Application to coherent receiver design*. NY: Wiley, 1976.

[49]   H. Meyr and G. Ascheid, *Synchronization in digital communications*. NY: Wiley, 1990.

[50]   A. Mileant and S. Hinedi, "Lock detection in Costas loops," *IEEE Trans. Commun.*, vol. 40, no. 3, pp. 480-483, Mar. 1992.

[51]   A. Mileant and S. Hinedi, "On the effects of phase jitter on QPSK lock detection," *IEEE Trans. Commun.*, vol. 41, no. 7, pp. 1043-1046, Jul. 1993.

[52]   K. Yi, et al., "A new lock detection algorithm for QPSK digital demodulator," in *Proc. Seventh IEEE International Symposium on Personal, Indoor and Mobile Radio Communications (PIMRC'96)*, Taipei, Taiwan, Oct. 15-18, 1996, pp. 848-852.

[53]   L. Kyung Ha, J. Seung Chul, and C. Hyung Jin, "A novel digital lock detector for QPSK receiver," *IEEE Trans. Commun.*, vol. 46, no. 6, pp. 750-753, Jun. 1998.

[54]   M. R. Spiegel, *Mathematical handbook of formulas and tables*. NY: McGraw-Hill, 1968.

[55]   R. N. McDonough and A. D. Whalen, *Detection of signals in noise*, 2nd ed. CA: Academic Press, 1995.

[56]   H. Gudbjartsson and S. Patz, "The Rician distribution of noisy MRI data," *Magnetic Resonance in Medicine*, vol. 34, no. 6, pp. 910-914, Dec. 1995.

[57]   J. Sijbers, "Signal and Noise Estimation from Magnetic Resonance Images," Ph.D. Thesis, University of Antwerp, Antwerp, Belgium, 1998.

[58]   B. T. Kopp and W. P. Osborne, "Phase jitter in MPSK carrier tracking loops: analytical, simulation and laboratory results," *IEEE Trans. Commun.*, vol. 45, no. 11, pp. 1385-1388, Nov. 1997.

[59]   N. C. Beaulieu, A. S. Toms, and D. R. Pauluzzi, "Comparison of four SNR estimators for QPSK modulations," *IEEE Commun. Letters*, vol. 4, no. 2, pp. 43-45, Feb. 2000.

[60]   D. R. Pauluzzi and N. C. Beaulieu, "A comparison of SNR estimation techniques for the AWGN channel," *IEEE Trans. Commun.*, vol. 48, no. 10, pp. 1681-1691, Oct. 2000.

[61]   D. R. Pauluzzi and N. C. Beaulieu, "A comparison of SNR estimation techniques in the AWGN channel," in *Proc. IEEE Pacific Rim Conference on Communications, Computers, and Signal Processing*, May 17-19, 1995, pp. 36-39.

[62]   N. Celandroni, E. Ferro, and F. Potorti, "Quality estimation of PSK modulated signals," *IEEE Commun. Mag.*, vol. 35, no. 7, pp. 50-55, Jul. 1997.

[63]   N. Celandroni and S. T. Rizzo, "Detection of errors recovered by decoders for signal quality estimation on rain-faded AWGN satellite channels," *IEEE Trans. Commun.*, vol. 46, no. 4, pp. 446-449, Apr. 1998.

[64]   N. Celandroni and S. T. Rizzo, "Corrections to 'Detection of Errors Recovered by Decoders for Signal Quality Estimation on Rain Faded AWGN Satellite Channels'," *IEEE Trans. Commun.*, vol. 47, no. 5, pp. 784-784, May 1999.

[65]   G. Ping and C. Tepedelenlioglu, "SNR estimation for nonconstant modulus constellations," *IEEE Trans. Signal Proc.*, vol. 53, no. 3, pp. 865-870, Mar. 2005.

[66]   M. K. Simon and S. Dolinar, "Improving SNR estimation for autonomous receivers," *IEEE Trans. Commun.*, vol. 53, no. 6, pp. 1063-1073, Jun. 2005.

[67]   A. Ramesh, A. Chockalingam, and L. B. Milstein, "SNR estimation in Nakagami-m fading with diversity combining and its application to turbo decoding," *IEEE Trans. Commun.*, vol. 50, no. 11, pp. 1719-1724, Nov. 2002.

[68]   R. Matzner, F. Engleberger, and R. Siewert, "Analysis and Design of a Blind Statistical SNR Estimator," in *Proc. Audio Engineering Society (AES) 102nd Convention*, Munich, Germany, Mar. 22-25, 1997.

[69]   A. Wiesel, J. Goldberg, and H. Messer, "Non-data-aided signal-to-noise-ratio estimation," in *Proc. IEEE International Conference on Communications (ICC 2002)*. Apr. 28 - May 2, 2002, pp. I:197-201.

[70]   A. Wiesel, J. Goldberg, and H. Messer-Yaron, "SNR estimation in time-varying fading channels," *IEEE Trans. Commun.*, vol. 54, no. 5, pp. 841-848, May 2006.

[71]   X. Hua and Z. Hui, "The simple SNR estimation algorithms for MPSK signals," in *Proc. 7th International Conference on Signal Processing (ICSP '04)*, Aug. 31 - Sept. 4 2004, pp. II:1781-1785.

[72]   R. Guangliang, C. Yilin, and Z. Hui, "A new SNR's estimator for QPSK Modulations in an AWGN channel," *IEEE Trans. on Circuits and Systems II: Express Briefs*, vol. 52, no. 6, pp. 336-338, Jun. 2005.

[73]   C. F. Mecklenbrauker and S. Paul, "On estimating the signal to noise ratio from BPSK signals," in *Proc. IEEE Intl. Conf. on Acoustics, Speech, and Signal Processing (ICASSP '05)*, Mar. 18-23, 2005, pp. IV:65-68.

[74]   L. Bin, R. DiFazio, and A. Zeira, "A low bias algorithm to estimate negative SNRs in an AWGN channel," *IEEE Commun. Letters*, vol. 6, no. 11, pp. 469-471, Nov. 2002.

[75]   E. Biglieri, J. Proakis, and S. Shamai, "Fading channels: information-theoretic and communications aspects," *IEEE Trans. Info. Theory*, vol. 44, no. 6, pp. 2619-2692, Jun. 1998.

[76]   M. K. Simon and M.-S. Alouini, *Digital communication over fading channels: a unified approach to performance analysis*. NY: Wiley, 2000.

[77]   M. K. Simon and M. Alouini, "A unified approach to the performance analysis of digital communication over generalized fading channels," *Proc. IEEE*, vol. 86, no. 9, pp. 1860-1877, Sep. 1998.

[78]   B. Sklar, "Rayleigh fading channels in mobile digital communication systems. I. Characterization," *IEEE Comm. Mag.*, vol. 35, no. 9, pp. 136-146, Sep. 1997.

[79]   B. Sklar, "Rayleigh Fading Channels in Mobile Digital Communication Systems Part II: Mitigation," *IEEE Comm. Mag.*, vol. 35, no. 9, pp. 148-155, Sep. 1997.

[80]   A. J. Viterbi and A. M. Viterbi, "Nonlinear estimation of PSK-modulated carrier phase with application to burst digital transmission," *IEEE Trans. Info. Theory*, vol. 29, no. 4, pp. 543-551, Jul. 1983.

[81]   R. Hamila, J. Vesma, and M. Renfors, "Polynomial-based maximum-likelihood technique for synchronization in digital receivers," *IEEE Trans. Circuits and Systems II*, vol. 49, no. 8, pp. 567-576, Aug. 2002.

[82]   D. Taich and I. Bar-David, "Maximum-likelihood estimation of phase and frequency of MPSK signals," *IEEE Trans. Info. Theory*, vol. 45, no. 7, pp. 2652-2655, Jul. 1999.

[83]   R. Hamila, "Synchronization and Multipath Delay Estimation Algorithms for Digital Receivers," Ph.D. Thesis, Tampere University of Technology, Tampere, Finland, 2002.

[84]   N. Noels, et al., "Carrier phase and frequency estimation for pilot-symbol assisted transmission: bounds and algorithms," *IEEE Trans. Signal Proc.*, vol. 53, no. 12, pp. 4578-4587, Dec. 2005.

[85]   M. L. Boucheret, et al., "A new algorithm for nonlinear estimation of PSK-modulated carrier phase," in *Proc. 3rd European Conference on Satellite Communications*, Manchester, UK, 1993, pp. 155-159.

[86]   W. G. Cowley, "Phase and frequency estimation for PSK packets: bounds and algorithms," *IEEE Trans. Commun.*, vol. 44, no. 1, pp. 26-28, Jan. 1996.

[87]   M. Moeneclaey and G. de Jonghe, "ML-oriented NDA carrier synchronization for general rotationally symmetric signal constellations," *IEEE Trans. Commun.*, vol. 42, no. 8, pp. 2531-2533, Aug. 1994.

[88]   N. A. D'Andrea, U. Mengali, and R. Reggiannini, "Comparison of carrier recovery methods for narrow-band polyphase shift keyed signals," in *Proc. GLOBECOM '88*, Hollywood, FL, USA, 1988, pp. 1474-1478.

[89]   J. B. Anderson, T. Aulin, and C.-E. Sundberg, *Digital phase modulation*. NY: Plenum Press, 1986.

[90]   E. A. Lee and D. G. Messerschmitt, *Digital communication*, 2nd ed. Boston: Kluwer Academic Publishers, 1994.

[91]   H. C. Osborne, "A generalized 'polarity-type' Costas loop for tracking MPSK signals," *IEEE Trans. Commun.*, vol. 30, no. 10, pp. 2289-2296, Oct. 1982.

[92]   S. A. Butman and J. R. Lesh, "The effects of bandpass limiters on n-phase tracking systems," *IEEE Trans. Commun.*, vol. 25, no. 6, pp. 569-576, Jun. 1977.

[93]   B. T. Kopp, "An analysis of carrier phase jitter in an MPSK receiver Utilizing MAP estimation," Ph.D. Thesis, New Mexico State University, Las Cruces, NM, 1994.

[94]   C. Dick, F. Harris, and M. Rice, "Synchronization in software radios. Carrier and timing recovery using FPGAs," in *Proc. 2000 IEEE Symposium on Field-Programmable Custom Computing Machines*, Napa Valley, CA, USA, Apr. 17-19, 2000, pp. 195-204.

[95]   L. Franks, "Carrier and Bit Synchronization in Data Communication - A Tutorial Review," *IEEE Trans. Commun.*, vol. 28, no. 8, pp. 1107-1121, Aug. 1980.

[96]   R. Hayashi, F. Ishizu, and K. Murakami, "A delta-sigma baseband phase detector realizing AGC-free PSK and FSK receivers," in *Proc. 13th IEEE International Symposium on Personal, Indoor and Mobile Radio Communications (PIMRC)*, Sept. 15-18, 2002, pp. 2382-2388.

[97] W. P. Osborne and B. T. Kopp, "Synchronization in M-PSK modems," in *Proc. ICC '92*, Chicago, IL, USA, 1992, pp. 1436-1440.

[98] R. De Gaudenzi, T. Garde, and V. Vanghi, "Performance analysis of decision-directed maximum-likelihood phase estimators for M-PSK modulated signals," *IEEE Trans. Commun.*, vol. 43, no. 12, pp. 3090-3100, Dec. 1995.

[99] G. Ascheid and H. Meyr, "Cycle Slips in Phase-Locked Loops: A Tutorial Survey," *IEEE Trans. Commun.*, vol. 30, no. 10, pp. 2228-2241, Oct. 1982.

[100] W. Lindsey and C. Chak, "Performance Measures for Phase-Locked Loops - A Tutorial," *IEEE Trans. Commun.*, vol. 30, no. 10, pp. 2224-2227, Oct. 1982.

[101] M. C. Jeruchim, P. Balaban, and K. S. Shankugan, *Simulation of Communication Systems*. New York: Plenum Publishing Corporation, 1992.

[102] J. G. Proakis and M. Salehi, *Contemporary Communication Systems Using Matlab*. Pacific Grove, CA: Brooks/Cole, 2000.

[103] R. E. Best, *Phase-locked loops: theory, design, and applications*, 2nd ed. NY: McGraw-Hill, 1993.

[104] Y. Linn, "An Optimal Adaptive M-PSK Carrier Phase Detector Suitable for Fixed-Point Hardware Implementation within FPGAs and ASICs," in *Proc. IEEE 2006 Workshop on Signal Processing Systems (SiPS'06)*, Banff, AB, Canada, Oct. 2-4, 2006, pp. 238-243.

[105] K. Steiglitz and L. McBride, "A technique for the identification of linear systems," *IEEE Trans. Automatic Control*, vol. 10, no. 4, pp. 461-464, Oct. 1965.

[106] G. Engeln-Müllges and F. Uhlig, *Numerical Algorithms with C*. Berlin: Springer, 1996.

[107] J. W. Craig, "A new, simple and exact result for calculating the probability of error for two-dimensional signal constellations," in *Proc. MILCOM'91*, 4-7 Nov., 1991, pp. 571-575.

[108] S. Hyundong and L. Jae Hong, "On the error probability of binary and M-ary signals in Nakagami-m fading channels," *IEEE Trans. Commun.*, vol. 52, no. 4, pp. 536-539, Apr. 2004.

[109] J. H. Roberts, *Angle modulation: the theory of system assessment*. Stevenage, UK: Peter Peregrinus Ltd., 1977.

[110] R. F. Pawula, S. O. Rice, and J. H. Roberts, "Distribution of the phase angle between two vectors perturbed by Gaussian noise," *IEEE Trans. Commun.*, vol. 30, no. 8, pp. 1828-1841, Aug. 1982.

[111] R. F. Pawula, "Distribution of the phase angle between two vectors perturbed by Gaussian noise II," *IEEE Trans. Veh. Technol.*, vol. 50, no. 2, pp. 576-583, Mar. 2001.

[112] R. Pawula, "On M-ary DPSK Transmission Over Terrestrial and Satellite Channels," *IEEE Trans. Commun.*, vol. 32, no. 7, pp. 752-761, Jul. 1984.

[113] N. Blachman, "The Effect of Phase Error on DPSK Error Probability," *IEEE Trans. Commun.*, vol. 29, no. 3, pp. 364-365, Mar. 1981.

[114] Xilinx Inc., "Virtex Series FPGAs," at http://www.xilinx.com/products/silicon_solutions/fpgas/virtex/index.htm, accessed Nov. 2006

[115] Altera Inc., "Altera Product Catalog Jan. 2006," at http://www.altera.com/literature/lit-index.html, accessed Nov. 2006

[116] A. Annamalai and C. Tellambura, "Error rates for Nakagami-m fading multichannel reception of binary and M-ary signals," *IEEE Trans. Commun.*, vol. 49, no. 1, pp. 58-68, Jan. 2001.

[117] K. L. Chung, *A course in probability theory*. NY: Harcourt, 1968.

[118] R. F. Pawula, "Generic error probabilities," *IEEE Trans. Commun.*, vol. 47, no. 5, pp. 697-702, May 1999.

[119] J. R. Barry, E. A. Lee, and D. G. Messerschmitt, *Digital communication*, 3rd ed. Boston: Kluwer, 2004.

[120] G. Karam, V. Paxal, and M. Moeneclaey, "Lock detectors for timing recovery," in *Proc. ICC '96.*, Dallas, TX, USA, 1996, pp. 1281-1285.

[121] Y. Linn, "A Hardware Method for Real-Time SNR Estimation for M-PSK using a Symbol Synchronization Lock Metric," in *Proc. 9th Canadian Workshop on Information Theory*, Montréal, QC, Canada, Jun. 5-8, 2005, pp. 247-251.

[122] Y. Linn, "A symbol synchronization lock detector and SNR estimator for QPSK, with application to BPSK," in *Proc. 3rd IASTED International Conference on Wireless and Optical Communications (IASTED WOC'03)*, Banff, AB, Canada, Jul. 14-16, 2003, pp. 506-514.

[123] Y. Linn, "A self-normalizing symbol synchronization lock detector for QPSK and BPSK," *IEEE Trans. Wireless Commun.*, vol. 5, no. 2, pp. 347-353, Feb. 2006.

[124] K.-P. Ho, *Phase-modulated optical communication systems*. NY: Springer, 2005.

APPENDICES

Appendix A Closed-Form Expressions for $f_M(\chi)$

In (30) we presented a formula for the computation of the lock detector expectation, repeated here for convenience:

$$f_M(\chi) = \int_{-\pi}^{\pi} \cos(M\Delta\phi) \cdot p_R(\Delta\phi|\chi) \cdot d\Delta\phi \qquad (222)$$

which is based upon the probability distribution of (29), also repeated for convenience:

$$p_R(\Delta\phi|\chi) \triangleq p(\Delta\phi_n = \Delta\phi | E_S/N_0 = \chi)$$

$$= \frac{\exp(-\chi)}{2\pi} \times \left[1 + \sqrt{2\chi} \cos(\Delta\phi) \exp(\chi \cdot \cos^2(\Delta\phi)) \cdot \int_{-\infty}^{\cos(\Delta\phi)\sqrt{2\chi}} e^{-y^2/2} dy \right] \qquad (223)$$

While at first glance due to the complicated nature of (223) closed-form formulas for (222) may seem unattainable, in fact using Fourier analysis we can reach formulas which are given entirely as finite sums of elementary functions.

We begin by noting that the domain of $p_R(\Delta\phi|\chi)$ is $[-\pi, \pi]$ and that therefore its periodic continuation can be represented as a Fourier series. Such an analysis has been conducted in [124 App. 4A]. The Fourier series coefficients for $p_R(\Delta\phi|\chi)$ are given by [124 eq. 4.A.9]:

$$c_m \triangleq \int_{-\pi}^{\pi} p_R(\Delta\phi|\chi) \exp(j \cdot m \cdot \Delta\phi) \cdot d\Delta\phi$$

$$= \int_{-\pi}^{\pi} p_R(\Delta\phi|\chi) \cos(m \cdot \Delta\phi) \cdot d\Delta\phi + j \cdot \int_{-\pi}^{\pi} p_R(\Delta\phi|\chi) \sin(m \cdot \Delta\phi) \cdot d\Delta\phi \qquad (224)$$

This appendix was presented in part in Y. Linn, "Simple and Exact Closed-Form Expressions for the Expectation of the Linn-Peleg M-PSK Lock Detector," in *Proc. 2007 IEEE Pacific Rim Conference on Communications, Computers and Signal Processing (PACRIM'07)*, Victoria, BC, Canada, Aug. 22-24, 2007, pp. 102-104.

Now we make use of the fact that $p_R\left(\Delta\phi\middle|\chi\right)$ is an even function of $\Delta\phi$ to conclude that the imaginary part of (224) vanishes, so that:

$$c_m = \int_{-\pi}^{\pi} p_R\left(\Delta\phi\middle|\chi\right)\cos(m\cdot\Delta\phi)\cdot d\Delta\phi \tag{225}$$

Comparing (222) to (225), we find that

$$f_M\left(\chi\right)=c_M \tag{226}$$

Hence, if we can find a closed-form formula for c_m then we can find a closed-form expression for $f_M(\chi)$. Fortunately, the coefficients c_m have been investigated in the literature, and a formula for them is given by ([124 eq. 4.A.11], with conversion to the appropriate notations used here) :

$$c_m = \frac{\sqrt{\pi\cdot\chi}}{2}\cdot\exp\left(\frac{-\chi}{2}\right)\left[I_{\frac{m-1}{2}}\left(\frac{\chi}{2}\right)+I_{\frac{m+1}{2}}\left(\frac{\chi}{2}\right)\right] \tag{227}$$

where $I_k\left(\bullet\right)$ is the k-th order modified Bessel function of the first kind (see [54 Chap. 24]). Therefore from (226):

$$f_M\left(\chi\right)=\frac{\sqrt{\pi\cdot\chi}}{2}\cdot\exp\left(\frac{-\chi}{2}\right)\left[I_{\frac{M-1}{2}}\left(\frac{\chi}{2}\right)+I_{\frac{M+1}{2}}\left(\frac{\chi}{2}\right)\right] \tag{228}$$

Moreover, given that M is always an even number, we have that $\frac{M-1}{2}$ and $\frac{M+1}{2}$ are always half of an odd integer. Given this fact, we can express these modified Bessel functions in closed form, given the recursive relation [54 eq. (24.46)]:

$$I_{n+1}(x) = I_{n-1}(x) - \frac{2n}{x}I_n(x) \tag{229}$$

Along with [54 eq. (24.58)]:

$$I_{1/2}(x) = \sqrt{\frac{2}{\pi x}}\sinh(x) \tag{230}$$

and [54 eq. (24.60)]:

$$I_{3/2}(x) = \sqrt{\frac{2}{\pi x}}\left(\cosh(x) - \frac{\sinh(x)}{x}\right) \tag{231}$$

For example, using (230) and (231) in (228) we find that for $M = 2$:

$$f_2(\chi) = \frac{\sqrt{\pi \cdot \chi}}{2} \cdot \exp\left(\frac{-\chi}{2}\right)\left[I_{1/2}\left(\frac{\chi}{2}\right) + I_{3/2}\left(\frac{\chi}{2}\right)\right]$$

$$= \frac{\sqrt{\pi \cdot \chi}}{2} \cdot \exp\left(\frac{-\chi}{2}\right)\left[\left(\sqrt{\frac{2}{\pi\left(\frac{\chi}{2}\right)}}\sinh\left(\frac{\chi}{2}\right)\right) + \sqrt{\frac{2}{\pi\left(\frac{\chi}{2}\right)}}\left(\cosh\left(\frac{\chi}{2}\right) - \frac{\sinh\left(\frac{\chi}{2}\right)}{\frac{\chi}{2}}\right)\right]$$

$$= \frac{\sqrt{\pi \cdot \chi}}{2} \cdot \exp\left(\frac{-\chi}{2}\right)\sqrt{\frac{4}{\pi\chi}}\left[\sinh\left(\frac{\chi}{2}\right) + \cosh\left(\frac{\chi}{2}\right) - \frac{\sinh\left(\frac{\chi}{2}\right)}{\frac{\chi}{2}}\right]$$

(232)

$$= \exp\left(\frac{-\chi}{2}\right)\left[\exp\left(\frac{\chi}{2}\right) - \frac{\exp\left(\frac{\chi}{2}\right) - \exp\left(\frac{-\chi}{2}\right)}{\chi}\right] = \left[1 - \frac{1 - \exp(-\chi)}{\chi}\right] = 1 - \frac{1}{\chi} + \frac{\exp(-\chi)}{\chi}$$

Similarly, for $M = 4$ we have from [54 eq. (24.62)] (or, equivalently, using (229)-(231)) that:

$$I_{5/2}(x) = \sqrt{\frac{2}{\pi x}}\left(\left(\frac{3}{x^2}+1\right)\sinh(x) - \frac{3}{x}\cosh(x)\right)$$

(233)

and that therefore:

$$f_4(\chi) = \left(\frac{\sqrt{\pi \cdot \chi}}{2} \cdot \exp\left(\frac{-\chi}{2}\right)\left[I_{\frac{M-1}{2}}\left(\frac{\chi}{2}\right) + I_{\frac{M+1}{2}}\left(\frac{\chi}{2}\right)\right]\right)_{M=4}$$

$$= \exp\left(\frac{-\chi}{2}\right)\left[\left(\cosh\left(\frac{\chi}{2}\right) - \frac{\sinh\left(\frac{\chi}{2}\right)}{\frac{\chi}{2}}\right) + \left(\frac{3}{\left(\frac{\chi}{2}\right)^2}+1\right)\sinh\left(\frac{\chi}{2}\right) - \frac{3}{\frac{\chi}{2}}\cosh\left(\frac{\chi}{2}\right)\right]$$

(234)

$$= \exp\left(\frac{-\chi}{2}\right)\left[\left(1 - \frac{6}{\chi}\right)\cosh\left(\frac{\chi}{2}\right) + \left(\frac{12}{\chi^2} - \frac{2}{\chi}+1\right)\sinh\left(\frac{\chi}{2}\right)\right]$$

$$= \frac{\exp\left(\frac{-\chi}{2}\right)}{2}\left[\left(1 - \frac{6}{\chi}\right)\left(\exp\left(\frac{\chi}{2}\right) + \exp\left(\frac{-\chi}{2}\right)\right) + \left(\frac{12}{\chi^2} - \frac{2}{\chi}+1\right)\left(\exp\left(\frac{\chi}{2}\right) - \exp\left(\frac{-\chi}{2}\right)\right)\right] = \left(1 - \frac{4}{\chi} + \frac{6}{\chi^2}\right) + \left(-\frac{2}{\chi} - \frac{6}{\chi^2}\right)\exp(-\chi)$$

Using a similar procedure, we find that:

$$f_8(\chi) = \frac{\sqrt{\pi \cdot \chi}}{2} \cdot \exp\left(\frac{-\chi}{2}\right)\left[I_{\frac{7}{2}}\left(\frac{\chi}{2}\right) + I_{\frac{9}{2}}\left(\frac{\chi}{2}\right)\right]$$

$$= \left(1 - \frac{16}{\chi} + \frac{120}{\chi^2} - \frac{480}{\chi^3} + \frac{840}{\chi^4}\right) + \left(\frac{4}{\chi} - \frac{60}{\chi^2} - \frac{360}{\chi^3} - \frac{840}{\chi^4}\right)\exp(-\chi)$$

(235)

In general for even M we find that:

$$f_M(\chi)$$

$$= 1 + \frac{M}{2}\sum_{n=1}^{M/2}\frac{(-1)^n}{n!}\cdot\frac{(M/2+n-1)!}{(M/2-n)!\chi^n}$$

$$+ \exp(-\chi)\left[(-1)^{M/2+1}\sum_{n=1}^{M/2}\frac{1}{(n-1)!}\frac{(M/2+n-1)!}{(M/2-n)!\chi^n}\right]$$

(236)

$$= \frac{M}{2}\sum_{n=0}^{M/2}\frac{(-1)^n}{n!}\cdot\frac{(M/2+n-1)!}{(M/2-n)!\chi^n}$$

$$+ \exp(-\chi)\left[(-1)^{M/2+1}\sum_{n=1}^{M/2}\frac{1}{(n-1)!}\frac{(M/2+n-1)!}{(M/2-n)!\chi^n}\right]$$

(see also [92] for a different problem in which the factors of (227) arose. There, a different simplification approach was used but the same result as (236) was reached, hence providing additional validation for these derivations).

It is emphasized that (236) is a summation of a <u>finite</u> number of terms which are composed entirely of exponentials and polynomials (as exemplified in (232) for $M = 2$). Hence, eq. (236) is very conducive to easy computation via computer algorithms (and even manually for small M).

Appendix B Closed-Form Expressions for $\overline{f}_M(\overline{\chi})$ in the Presence of Nakagami-*m* Fading

Depending upon the fading probability distribution, we can sometimes use the results of Appendix A in order to obtain closed-form expressions for $\overline{f}_M(\overline{\chi})$ in the presence of fading. A case in point is the Nakagami-*m* distribution. From Table 2 we have in this case that:

$$p_{Nak-m}(\chi|\overline{\chi}) \triangleq \frac{m^m \chi^{m-1}}{\overline{\chi}^m \Gamma(m)} \exp\left(\frac{-m\chi}{\overline{\chi}}\right) \tag{237}$$

In order to facilitate the ensuing computations it is advantageous to compute some preliminary definite integrals. We do this using the integral [54 eq. (15.76)] which is:

$$\int_0^\infty \tau^n \exp(-a\tau)d\tau = \frac{\Gamma(n+1)}{a^{n+1}} \tag{238}$$

and we use this result to define the following function:

$$\Upsilon_{k,\ell,m}(\overline{\chi}) \triangleq \int_0^\infty y^{-k} \exp(-\ell y) \cdot p_{Nak-m}(y|\overline{\chi})dy$$

$$= \int_0^\infty y^{-k} \exp(-\ell y) \frac{m^m y^{m-1}}{\overline{\chi}^m \Gamma(m)} \exp\left(\frac{-m}{\overline{\chi}}y\right)dy \tag{239}$$

$$= \frac{m^m}{\overline{\chi}^m \Gamma(m)} \int_0^\infty y^{m-k-1} \exp\left(-\left(\ell + \frac{m}{\overline{\chi}}\right)y\right)dy$$

This appendix was presented in part in Y. Linn, "Simple and Exact Closed-Form Expressions for the Expectation of the Linn-Peleg M-PSK Lock Detector," in *Proc. 2007 IEEE Pacific Rim Conference on Communications, Computers and Signal Processing (PACRIM'07)*, Victoria, BC, Canada, Aug. 22-24, 2007, pp. 102-104.

$$= \frac{m^m}{\overline{\chi}^m \Gamma(m)} \cdot \frac{\Gamma(m-k)}{\left(\ell + \dfrac{m}{\overline{\chi}}\right)^{m-k}} = \frac{m^m}{\overline{\chi}^m \Gamma(m)} \cdot \frac{\Gamma(m-k)}{\left(\dfrac{\ell\overline{\chi}+m}{\overline{\chi}}\right)^{m-k}}$$

$$= \frac{\overline{\chi}^{-k} m^m \cdot \Gamma(m-k)}{\left(\ell\overline{\chi}+m\right)^{m-k} \Gamma(m)}$$

where $k, \ell \geq 0$ and $m \geq 0.5$. A minor difficulty may arise when using (239): when $m-k$ is a non-positive integer, then $\Gamma(m-k)$ tends to positive or negative infinity (see [54 Fig. 16.1]). This problem is elegantly solved by substituting $\tilde{m} = m + \varepsilon$ instead of m in (239), where ε is a very small number, i.e. $0 < \varepsilon \ll m$, such as $\varepsilon = 0.001 \cdot m$. Doing so will have a negligible effect upon the results while avoiding the singularity of $\Gamma(m-k)$. Hence we define the following function:

$$\tilde{\Upsilon}_{k,\ell,m}\left(\overline{\chi}\right) = \begin{cases} \Upsilon_{k,\ell,m}\left(\overline{\chi}\right) & \text{m-k} \notin \{0,-1,-2,-3,\ldots\} \\ \Upsilon_{k,\ell,\tilde{m}}\left(\overline{\chi}\right) & \text{m-k} \in \{0,-1,-2,-3,\ldots\} \end{cases} \tag{240}$$

where $k, \ell \geq 0$ and $m \geq 0.5$, and $\tilde{m} \triangleq m + m/1000$

From (58) we have that $\overline{f}_M\left(\overline{\chi}\right) \triangleq \int_0^\infty f_M\left(\chi\right) p_{Nak-m}\left(\chi \mid \overline{\chi}\right) d\chi$. We can now use (239)-(240) in conjunction with the results of Appendix A in order to arrive at closed-form expressions for $\overline{f}_M\left(\overline{\chi}\right)$ for the case of Nakagami-m fading. For example, for $M = 2$ we have from (232), (239) and (240) that:

$$\overline{f}_2(\overline{\chi}) = \tilde{\Upsilon}_{0,0,m}(\overline{\chi}) - \tilde{\Upsilon}_{1,0,m}(\overline{\chi}) + \tilde{\Upsilon}_{1,1,m}(\overline{\chi})$$

$$= 1 - \frac{\overline{\chi}^{-1} \cdot \Gamma(m-1)}{m^{-1} \cdot \Gamma(m)} + \frac{\overline{\chi}^{-1} m^m \cdot \Gamma(m-1)}{\left(\overline{\chi}+m\right)^{m-1} \Gamma(m)} \tag{241}$$

(in (241) we assume that $m \neq 1$. If $m = 1$ we use $m = 1.001$ in (241), as per (240)). Similarly, for $M = 4$ we have from (234), (239) and (240), we have:

$$\overline{f}_4(\overline{\chi}) = \tilde{\Upsilon}_{0,0,m}(\overline{\chi}) - 4\cdot\tilde{\Upsilon}_{1,0,m}(\overline{\chi}) + 6\cdot\tilde{\Upsilon}_{2,0,m}(\overline{\chi})$$
$$-2\cdot\tilde{\Upsilon}_{1,1,m}(\overline{\chi}) - 6\cdot\tilde{\Upsilon}_{2,1,m}(\overline{\chi})$$
$$=1-4\cdot\frac{\overline{\chi}^{-1}\cdot\Gamma(m-1)}{m^{-1}\cdot\Gamma(m)} + 6\cdot\frac{\overline{\chi}^{-2}\cdot\Gamma(m-2)}{m^{-2}\cdot\Gamma(m)} \tag{242}$$
$$-2\cdot\frac{\overline{\chi}^{-1}m^m\cdot\Gamma(m-1)}{(\overline{\chi}+m)^{m-1}\Gamma(m)} - 6\cdot\frac{\overline{\chi}^{-2}m^m\cdot\Gamma(m-2)}{(\overline{\chi}+m)^{m-2}\Gamma(m)}$$

(in (241) we assume that $m \neq 1$ and $m \neq 2$. If $m = 1$ or $m = 2$ we use $m = 1.001$ or $m = 2.002$, respectively, in (242), as per (240)). In the general case, for even M we have from (236), (239) and (240) that:

$$\overline{f}_M(\overline{\chi}) = \tilde{\Upsilon}_{0,0,m}(\overline{\chi}) + \frac{M}{2}\sum_{n=1}^{M/2}\frac{(-1)^n}{n!}\cdot\frac{(M/2+n-1)!}{(M/2-n)!}\cdot\tilde{\Upsilon}_{n,0,m}(\overline{\chi})$$
$$+\left[(-1)^{M/2+1}\sum_{n=1}^{M/2}\frac{1}{(n-1)!}\frac{(M/2+n-1)!}{(M/2-n)!}\tilde{\Upsilon}_{n,1,m}(\overline{\chi})\right] \tag{243}$$

It is emphasized that (243) is a finite sum of terms which are composed of ratios of finite polynomials and gamma functions, and is very easy to compute using various numerical computation packages such as Matlab. Furthermore, we note that the Rayleigh and one-sided Gaussian fading distributions are particular cases of the Nakagami-m distribution (with m=1 and m=0.5 respectively (see [77 Table 2])), so closed-form expressions for these fading distributions can be obtained from (243) as well.

Appendix C Closed-Form Expressions for $f_M^D(\chi)$

Through use of Fourier analysis, closed-form expressions may be attained for $f_M^D(\chi)$. To do so, we first re-define for convenience the density function $p_D(\Delta\phi^D \mid \chi)$ over the interval $[-\pi,\pi]$ (this is possible because the true phase is in the interval $[-\pi,\pi]$, as noted in Sec. 4.12). We name this distribution $\tilde{p}_D(\Delta\tilde{\phi}^D \mid \chi)$ (where $\Delta\tilde{\phi}^D \in [-\pi,\pi]$), and it is given by [19 eq. (7.3), p. 441]:

$$\tilde{p}_D(\Delta\tilde{\phi}^D \mid \chi)$$
$$= \tfrac{1}{2\pi} \int_0^{\pi/2} (\sin\tau)\cdot(1+\chi(1+\cos\Delta\tilde{\phi}^D \sin\tau))e^{-\chi(1-\cos\Delta\tilde{\phi}^D \sin\tau)}d\tau \qquad (244)$$

With this definition of $\tilde{p}_D$ we find from (193) that (assuming for simplicity $\Delta\omega = 0$):

$$f_M^D(\chi) = \int_{-\pi}^{\pi} \cos(M\cdot\Delta\tilde{\phi}^D)\tilde{p}_D(\Delta\tilde{\phi}^D \mid \chi)d(\Delta\tilde{\phi}^D) \qquad (245)$$

(Note that the limits of the integral in (245) are $[-\pi,\pi]$). Now, because the domain of $\tilde{p}_D$ is finite, the periodic extension of $\tilde{p}_D$ can be expressed as a Fourier series, i.e.

$\tilde{p}_D(\Delta\tilde{\phi}^D \mid \chi) = \tfrac{1}{2\pi}\sum_{m=-\infty}^{\infty} c_m^D \exp(-jm\Delta\tilde{\phi}^D)$ where, using the fact that $\tilde{p}_D$ is even:

$$c_m^D = \int_{-\pi}^{\pi} \cos(m\cdot\Delta\tilde{\phi}^D)\tilde{p}_D(\Delta\tilde{\phi}^D \mid \chi)d(\Delta\tilde{\phi}^D) \qquad (246)$$

(see [124 App. 4A]). From comparing (246) to (245), we see that $f_M^D(\chi) = c_m^D\big|_{m=M}$. The coefficients c_m^D were computed in [124 App. 4A], from which it follows that (see [124 eq. (4.A.18)]):

$$f_M^D(\chi) = c_m^D\big|_{m=M} = \frac{\pi\chi}{4}\cdot e^{-\chi}\left[I_{\frac{M-1}{2}}\left(\frac{\chi}{2}\right)+I_{\frac{M+1}{2}}\left(\frac{\chi}{2}\right)\right]^2 \qquad (247)$$

where $I_k(\bullet)$ is the k-th order modified Bessel function of the first kind (see [54 Chap. 24]).

Quick inspection of (247) and (227) shows that we have $c_m^D = (c_m)^2$, which in turn means that $f_M^D(\chi) = (f_M(\chi))^2$. Hence, from (236) we have:

$$
f_M^D(\chi) = \left(
\begin{array}{l}
\dfrac{M}{2} \displaystyle\sum_{n=0}^{M/2} \dfrac{(-1)^n}{n!} \cdot \dfrac{(M/2+n-1)!}{(M/2-n)!\chi^n} \\[4mm]
+ \exp(-\chi)\left[(-1)^{M/2+1} \displaystyle\sum_{n=1}^{M/2} \dfrac{1}{(n-1)!} \dfrac{(M/2+n-1)!}{(M/2-n)!\chi^n} \right]
\end{array}
\right)^2
\tag{248}
$$

expanding, we have:

$$
f_M^D(\chi) = \left(
\begin{array}{l}
\dfrac{M^2}{4} \left(
\begin{array}{l}
\displaystyle\sum_{n=0}^{M/2} \dfrac{(-1)^n}{n!} \cdot \dfrac{(M/2+n-1)!}{(M/2-n)!\chi^n} \\[4mm]
\times \displaystyle\sum_{k=0}^{M/2} \dfrac{(-1)^k}{k!} \cdot \dfrac{(M/2+k-1)!}{(M/2-k)!\chi^k}
\end{array}
\right) \\[10mm]
+ (-1)^{M/2+1} M \exp(-\chi) \\[3mm]
\times \left(
\begin{array}{l}
\displaystyle\sum_{n=0}^{M/2} \dfrac{(-1)^n}{n!} \cdot \dfrac{(M/2+n-1)!}{(M/2-n)!\chi^n} \\[4mm]
\times \left[\displaystyle\sum_{k=1}^{M/2} \dfrac{1}{(k-1)!} \dfrac{(M/2+k-1)!}{(M/2-k)!\chi^k} \right]
\end{array}
\right) \\[10mm]
+ \exp(-2\chi) \left[
\begin{array}{l}
\displaystyle\sum_{n=1}^{M/2} \dfrac{1}{(n-1)!} \dfrac{(M/2+n-1)!}{(M/2-n)!\chi^n} \\[4mm]
\times \displaystyle\sum_{k=1}^{M/2} \dfrac{1}{(k-1)!} \dfrac{(M/2+k-1)!}{(M/2-k)!\chi^k}
\end{array}
\right]
\end{array}
\right)
\tag{249}
$$

Simplifying:

$$f_M^D(\chi) = \frac{M^2}{4} \sum_{n=0}^{M/2} \sum_{k=0}^{M/2} \frac{(-1)^{n+k}}{n!k!} \cdot \frac{(M/2+n-1)!(M/2+k-1)!}{(M/2-n)!(M/2-k)!\chi^{n+k}}$$

$$+(-1)^{M/2+1} M \cdot e^{-\chi} \sum_{n=0}^{M/2} \sum_{k=1}^{M/2} \frac{(-1)^n}{n!(k-1)!} \frac{(M/2+n-1)!(M/2+k-1)!}{(M/2-n)!(M/2-k)!\chi^{n+k}} \quad (250)$$

$$+\exp(-2\chi) \sum_{n=1}^{M/2} \sum_{k=1}^{M/2} \frac{(M/2+n-1)!(M/2+k-1)!}{(n-1)!(k-1)!(M/2-n)!(M/2-k)!\chi^{n+k}}$$

Following this procedure, we arrive at the following expressions:

$$f_2^D(\chi) = 1 - 2\chi^{-1} + \chi^{-2} + 2\chi^{-1}e^{-\chi} - 2\chi^{-2}e^{-\chi} + \chi^{-2}e^{-2\chi} \quad (251)$$

$$f_4^D(\chi) = 1 - 8\chi^{-1} + 28\chi^{-2} - 48\chi^{-3} + 36\chi^{-4} - 4\chi^{-1}e^{-\chi} + 4\chi^{-2}e^{-\chi}$$
$$+ 4\chi^{-2}e^{-2\chi} + 24\chi^{-3}e^{-\chi} + 24\chi^{-3}e^{-2\chi} - 72\chi^{-4}e^{-\chi} + 36\chi^{-4}e^{-2\chi} \quad (252)$$

Expressions for $M > 4$ can also be found. However, since those expressions are very tedious and since they can be arrived at following the same procedure outlined above, they are omitted. Moreover, we note that the approximation (196) is extremely accurate for $M > 2$ (see Sec. 4.12 and Fig. 97), so as a practical issue use of those expressions is often unnecessary.

<u>Appendix D</u> Closed-Form Expressions for $\bar{f}_M^D(\bar{\chi})$ in the Presence of Nakagami-*m* Fading

Using (248)-(252) we can compute $\bar{f}_M^D(\bar{\chi})$ via (198). For certain fading distributions, this can lead to closed form expressions for $\bar{f}_M^D(\bar{\chi})$. A case in point is again the very important Nakagami-m distribution.

From (198), exact closed-form expressions for $\bar{f}_M^D(\bar{\chi})$ can be obtained for Nakagami-*m* fading for all M and all m through computation of the definite integral

$$\bar{f}_M^D(\bar{\chi}) = \int_0^\infty f_M^D(\chi) \frac{m^m \chi^{m-1}}{\bar{\chi}^m \Gamma(m)} \exp\left(\frac{-m\chi}{\bar{\chi}}\right) d\chi \tag{253}$$

by using (248)-(252) along with the formula for the definite integral $\int_0^\infty y^n \exp(-ay) dy = \Gamma(n+1)/a^{n+1}$ (see (238)) and the definition of $\tilde{\Upsilon}_{k,\ell,m}(\bar{\chi})$ (see (239)-(240)). We now set upon doing this.

Using (250) in (253) we have

$$\bar{f}_M^D(\bar{\chi}) = \frac{M^2}{4} \sum_{n=0}^{M/2} \sum_{k=0}^{M/2} \frac{(-1)^{n+k}}{n!k!} \cdot \frac{(M/2+n-1)!(M/2+k-1)!}{(M/2-n)!(M/2-k)!} \tilde{\Upsilon}_{n+k,0,m}(\bar{\chi})$$

$$+(-1)^{M/2+1} M \sum_{n=0}^{M/2} \sum_{k=1}^{M/2} \frac{(-1)^n}{n!(k-1)!} \frac{(M/2+n-1)!(M/2+k-1)!}{(M/2-n)!(M/2-k)!} \tilde{\Upsilon}_{n+k,1,m}(\bar{\chi}) \tag{254}$$

$$+\sum_{n=1}^{M/2} \sum_{k=1}^{M/2} \frac{(M/2+n-1)!(M/2+k-1)!}{(n-1)!(k-1)!(M/2-n)!(M/2-k)!} \tilde{\Upsilon}_{n+k,2,m}(\bar{\chi})$$

For example, for *M*=2 we have:

$$\bar{f}_2^{D}\left(\bar{\chi}\right) = \tilde{\Upsilon}_{0,0,m}\left(\bar{\chi}\right) - 2 \cdot \tilde{\Upsilon}_{1,0,m}\left(\bar{\chi}\right) + \tilde{\Upsilon}_{2,0,m}\left(\bar{\chi}\right)$$

$$+ 2 \cdot \tilde{\Upsilon}_{1,1,m}\left(\bar{\chi}\right) - 2 \cdot \tilde{\Upsilon}_{2,1,m}\left(\bar{\chi}\right) + \tilde{\Upsilon}_{2,2,m}\left(\bar{\chi}\right)$$

$$= 1 - 2 \frac{\bar{\chi}^{-1} \cdot \Gamma(m-1)}{m^{-1}\Gamma(m)} + \frac{\bar{\chi}^{-2} \cdot \Gamma(m-2)}{m^{-2}\Gamma(m)} \tag{255}$$

$$+ 2 \frac{\bar{\chi}^{-1} m^{m} \cdot \Gamma(m-1)}{\left(\bar{\chi}+m\right)^{m-1} \Gamma(m)} - 2 \frac{\bar{\chi}^{-2} m^{m} \cdot \Gamma(m-2)}{\left(\bar{\chi}+m\right)^{m-2} \Gamma(m)}$$

$$+ \frac{\bar{\chi}^{-2} m^{m} \cdot \Gamma(m-2)}{\left(2\bar{\chi}+m\right)^{m-2} \Gamma(m)}$$

(in (255) we assume that $m \neq 1$ and $m \neq 2$. If $m = 1$ or $m = 2$ we use $m = 1.001$ or $m = 2.002$, respectively, in (255), as per (240)). Expressions for $M > 2$ can be obtained in a similar straightforward manner, though the resulting expressions are extremely long. It is important to note that through the method presented here we can compute $\bar{f}_M^{D}(\bar{\chi})$ for Nakagami-m fading using only elementary functions and gamma functions, something that can be easily done using numerical computational packages such as Matlab. These exact expressions for $\bar{f}_M^{D}(\bar{\chi})$ were used in the computation of Fig. 98, where we see that the theoretical results agree completely with those obtained through simulations.

Appendix E Asymptotic Limits and Simulation Computations of the Cross-Correlation Coefficients $\rho_{n,k}$

In this appendix we discuss the cross-correlation coefficients used in Sec. 4.12.3, and

we prove that $\rho_{n,k} = \begin{cases} 1 & n = k \\ \rho_1(\chi) & |n-k| = 1 \\ 0 & |n-k| > 1 \end{cases}$ where $|\rho_1(\chi)| \le 0.3$ and $\rho_{n,k}$ is the cross-

correlation coefficients of $\left\{ x_{M,n}^D \right\}_{n=-\infty}^{\infty}$ defined as:

$$\rho_{n,k} \triangleq \frac{E[x_{M,n}^D x_{M,k}^D] - E[x_{M,n}^D]E[x_{M,k}^D]}{\sqrt{\operatorname{var}(x_{M,n}^D)\operatorname{var}(x_{M,k}^D)}} \tag{256}$$

Let us first assume that no fading is involved. We begin by noting that $x_{M,N}^D = \cos\left(M\Delta\phi_n^D\right)\cos\left(M\Delta\omega T\right) = \cos\left(M\left(\Delta\phi_n - \Delta\phi_{n-1}\right)\right)\cos\left(M\Delta\omega T\right)$. Since we assumed $\Delta\omega \ll 2\pi/(M \cdot T)$ then this simplifies to $x_{M,n}^D = \cos\left(M\left(\Delta\phi_n - \Delta\phi_{n-1}\right)\right)$. We note that the variables $\left\{\Delta\phi_n\right\}_{n=-\infty}^{\infty}$ are mutually independent. From this it immediately follows that $x_{M,n}^D$ and $x_{M,k}^D$ will be independent for all $|n-k| > 1$, which means that $\rho_{n,k} = 0$ for $|n-k| > 1$. Moreover, the fact that $\rho_{n,n} = 1$ is a fundamental result from probability theory which can also be verified by inspection of the definition of $\rho_{n,k}$. Thus, the only remaining issue is the characterization of $\rho_{n,k}$ for $|n-k| = 1$, namely the characterization of the function $\rho_1(\chi)$.

As for operation with fading, we note that since we assumed slow fading, we can assume that the SNR remains constant over two symbol intervals. Thus, *ipso facto*, we will have the same correlation coefficient $\rho_{n,k}$ for $|n-k| = 1$ as for the non-fading case. Regarding the cases $n = k$ and $|n-k| > 1$, the arguments that led to the conclusion that

$\rho_{n,n}=1$ and $\rho_{n,k}=0$ for $|n-k|>1$ remain valid in the presence of slow fading. Hence, we conclude that the coefficients $\rho_{n,k}$ are unaffected by slow fading.

Pursuant to the preceding analysis, we now engage upon characterizing $\rho_{n,k}$ for $|n-k|=1$, which is the only thing which remains in order to fully qualify these variables. This can be done through stochastic simulations, i.e. through computation of $\rho_1(\chi)$ (the notation we use for $\rho_{n,k}$ for $|n-k|=1$) using simulated sequences of $I(n)$ and $Q(n)$. This is shown in Fig. 125. As seen there, we indeed have $|\rho_1(\chi)| \le 0.3$ which was the assumption used in Sec. 4.12.3.

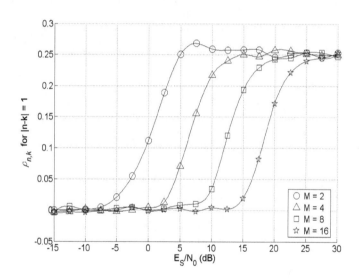

Fig. 125. $\rho_{n,k}$ for $|n-k|=1$ (also denoted as $\rho_1(\chi)$) as a function of the SNR ($=\chi$).

As seen in Fig. 125, we have $\rho_1(\chi) \xrightarrow{\chi \to 0} 0$ and $\rho_1(\chi) \xrightarrow{\chi \to \infty} 0.25$. We can actually justify these asymptotic values theoretically, an endeavour that we shall presently undertake.

- 259 -

First, let us take a loop at the case of $\chi \to 0$. In that limit, we have that there is no signal component in the values of $I(n)$ and $Q(n)$. It is then easy to show (and is also clear intuitively) that $\rho_{n,k}$ for $|n-k|=1$ tends to 0 (that is, lack of correlation).

Let us now take a look at $\chi \to \infty$ and prove that $\rho_1(\chi) \xrightarrow{\chi \to \infty} 0.25$. Assuming that $\chi \to \infty$, we have from the Taylor series expansion that

$$x_{M,n}^D = \cos\left(M\Delta\phi_n^D\right) \xrightarrow{\chi \to \infty} 1 - \frac{M^2\left(\Delta\phi_n^D\right)^2}{2}. \text{ Thus:}$$

$$\begin{aligned}
\text{var}\left(x_{M,n}^D\right) &= \text{var}\left(1 - \frac{M^2\left(\Delta\phi_n^D\right)^2}{2}\right) \\
&= E\left[\left(1 - \frac{M^2\left(\Delta\phi_n^D\right)^2}{2}\right)^2\right] - E^2\left[\left(1 - \frac{M^2\left(\Delta\phi_n^D\right)^2}{2}\right)\right] \\
&= \left(1 - M^2 E\left[\left(\Delta\phi_n^D\right)^2\right] + \frac{M^4 E\left[\left(\Delta\phi_n^D\right)^4\right]}{4}\right) - \left(1 - M^2 E\left[\left(\Delta\phi_n^D\right)^2\right] + \frac{M^2\left(E\left[\left(\Delta\phi_n^D\right)^2\right]\right)^2}{4}\right) \\
&= \frac{M^4}{4}\left(E\left[\left(\Delta\phi_n^D\right)^4\right] - \left(E\left[\left(\Delta\phi_n^D\right)^2\right]\right)^2\right)
\end{aligned}$$

(257)

Now, at high SNR we have $\Delta\phi_n^D \overset{\text{high SNR}}{\sim} N\left(0,\dfrac{1}{\chi}\right)$ so that $E\left[\left(\Delta\phi_n^D\right)^2\right] = \dfrac{1}{\chi}$ and $E\left[\left(\Delta\phi_n^D\right)^4\right] = \dfrac{3}{\chi^2}$. Thus from (257):

$$\text{var}\left(x_{M,n}^D\right) = \frac{M^4}{4}\left(\frac{3}{\chi^2} - \frac{1}{\chi^2}\right) = \frac{M^4}{2\chi^2}$$

(258)

We now turn our attention to the numerator of (256). We assume $k = n-1$ (a similar derivation will give the same result for $k = n+1$). Since $E[x_{M,n}^D] = E\left[\cos\left(M\Delta\phi_n^D\right)\right]$ and since $\Delta\phi_n^D \overset{\text{high SNR}}{\sim} N\left(0,1/\chi\right)$ we have (using $\int_0^\infty e^{-\alpha x^2}\cos(bx)dx = \frac{1}{2}\sqrt{\frac{\pi}{a}}e^{-b^2/(4a)}$ [54 eq. 15.73]):

$$E[x_{M,n}^D] = E[x_{M,n-1}^D]$$

$$\overset{\text{high SNR}}{=} \int_{-\infty}^{\infty} \cos(M\tau)\sqrt{\frac{\chi}{2\pi}} \exp\left(-\frac{\chi}{2}\tau^2\right) d\tau = \exp\left(-\frac{M^2}{2\chi}\right) \qquad (259)$$

Furthermore:

$$E[x_{M,n}^D x_{M,n-1}^D] = E\left[\cos\left(M\Delta\phi_n^D\right)\cos\left(M\Delta\phi_{n-1}^D\right)\right]$$

$$= E\left[\frac{1}{2}\cos\left(M\left(\Delta\phi_n^D + \Delta\phi_{n-1}^D\right)\right) + \frac{1}{2}\cos\left(M\left(\Delta\phi_n^D - \Delta\phi_{n-1}^D\right)\right)\right]$$

$$= \frac{1}{2}E\left[\begin{array}{l}\cos\left(M\left(\Delta\phi_n - \Delta\phi_{n-1} + \Delta\phi_{n-1} - \Delta\phi_{n-2}\right)\right) \\ + \cos\left(M\left(\Delta\phi_n - \Delta\phi_{n-1} - \Delta\phi_{n-1} + \Delta\phi_{n-2}\right)\right)\end{array}\right] \qquad (260)$$

$$= \frac{1}{2}\left(E\left[\cos\left(M\left(\Delta\phi_n - \Delta\phi_{n-2}\right)\right)\right] + E\left[\cos\left(M\left(\Delta\phi_n - 2\Delta\phi_{n-1} + \Delta\phi_{n-2}\right)\right)\right]\right)$$

Let us define $\psi \triangleq \Delta\phi_n - \Delta\phi_{n-2}$ and $\kappa = \Delta\phi_n - 2\Delta\phi_{n-1} + \Delta\phi_{n-2}$. At high SNR we have that

$$\Delta\phi_n \overset{\text{high SNR}}{\sim} N\left(0, 1/(2\chi)\right) \text{ for all } n. \text{ Hence, since } \{\Delta\phi_n\}_{n=-\infty}^{\infty} \text{ are mutually independent it}$$

is easy to show that we have:

$$\psi \sim N(0, 1/\chi) \qquad (261)$$

and:

$$\kappa \sim N(0, 3/\chi) \qquad (262)$$

Thus (using $\int_0^{\infty} e^{-ax^2}\cos(bx)dx = \frac{1}{2}\sqrt{\frac{\pi}{a}}e^{-b^2/(4a)}$ [54 eq. 15.73]):

$$E\left[\cos\left(M\left(\Delta\phi_n - \Delta\phi_{n-2}\right)\right)\right]$$

$$= \int_{-\infty}^{\infty} \cos(M\psi)\sqrt{\frac{\chi}{2\pi}} \exp\left(-\frac{\chi}{2}\psi^2\right) d\psi = \exp\left(-\frac{M^2}{2\chi}\right) \qquad (263)$$

and:

$$E\left[\cos\left(M\left(\Delta\phi_n - 2\Delta\phi_{n-1} + \Delta\phi_{n-2}\right)\right)\right]$$

$$= \int_{-\infty}^{\infty} \cos(M\kappa)\sqrt{\frac{\chi}{6\pi}} \exp\left(-\frac{\chi}{6}\kappa^2\right) d\kappa = \exp\left(-\frac{3M^2}{2\chi}\right) \qquad (264)$$

Substituting (263) and (264) into (260) we get:

$$E[x_{M,n}^{D} x_{M,n-1}^{D}] = \frac{1}{2}\left[\exp\left(-\frac{M^2}{2\chi}\right) + \exp\left(-\frac{3M^2}{2\chi}\right)\right] \tag{265}$$

Substituting (265), (259), and (258) into (256) we get:

$$\rho_{n,n-1} \triangleq \frac{\frac{1}{2}\left[\exp\left(-\frac{M^2}{2\chi}\right) + \exp\left(-\frac{3M^2}{2\chi}\right)\right] - \left(\exp\left(-\frac{M^2}{2\chi}\right)\right)^2}{\frac{M^4}{2\chi^2}} \tag{266}$$

It is convenient[21] to define $\xi \triangleq \frac{1}{\chi}$ and express (266) using this new variable. We have:

$$\rho_{n,n-1} \triangleq \frac{\frac{1}{2}\left[\exp\left(-\frac{M^2\xi}{2}\right) + \exp\left(-\frac{3M^2\xi}{2}\right)\right] - \left(\exp\left(-\frac{M^2\xi}{2}\right)\right)^2}{\frac{M^4\xi^2}{2}} \tag{267}$$

The limit $\chi \to \infty$ is equivalent to $\xi \to 0^+$. Taking this limit upon (267) and using L'hopital's rule we have:

$$\lim_{\xi \to 0^+} \rho_{n,n-1} = \lim_{\xi \to 0^+} \frac{\frac{d}{d\xi}\left(\frac{1}{2}\left[\exp\left(-\frac{M^2\xi}{2}\right) + \exp\left(-\frac{3M^2\xi}{2}\right)\right] - \left(\exp\left(-\frac{M^2\xi}{2}\right)\right)^2\right)}{\frac{d}{d\xi}\left(\frac{M^4\xi^2}{2}\right)}$$

$$= \lim_{\xi \to 0^+} \frac{\left(\begin{array}{c}\frac{1}{2}\left[-\frac{M^2}{2}\exp\left(-\frac{M^2\xi}{2}\right) - \frac{3M^2}{2}\exp\left(-\frac{3M^2\xi}{2}\right)\right] \\ -\left(2\left(-\frac{M^2}{2}\right)\exp\left(-\frac{M^2\xi}{2}\right)\exp\left(-\frac{M^2\xi}{2}\right)\right)\end{array}\right)}{\frac{M^4 \cdot 2\xi}{2}} \tag{268}$$

$$= \lim_{\xi \to 0^+} \frac{\frac{1}{2}\left[-\frac{M^2}{2}\exp\left(-\frac{M^2\xi}{2}\right) - \frac{3M^2}{2}\exp\left(-\frac{3M^2\xi}{2}\right)\right] + M^2\exp\left(-M^2\xi\right)}{M^4\xi}$$

Both the denominator and numerator still tend to 0 so we use L'hopital's rule again to yield:

---

[21] Note that this quantity has nothing to do with the phase detector self-noise that was used in Chapter 3 and which was written using a similar notation.

$$\lim_{\xi \to 0^+} \rho_{n,n-1} = \lim_{\xi \to 0^+} \frac{\dfrac{d}{d\xi}\left(\dfrac{1}{2}\left[-\dfrac{M^2}{2}\exp\left(-\dfrac{M^2\xi}{2}\right) \dfrac{3M^2}{2}\exp\left(-\dfrac{3M^2\xi}{2}\right)\right] \\ +M^2\exp\left(-M^2\xi\right)\right)}{\dfrac{d}{d\xi}\left(M^4\xi\right)}$$

$$= \lim_{\xi \to 0^+} \frac{\dfrac{1}{2}\left[\dfrac{M^4}{4}\exp\left(-\dfrac{M^2\xi}{2}\right) + \dfrac{9M^4}{4}\exp\left(-\dfrac{3M^2\xi}{2}\right)\right] - \left(M^4\exp\left(-M^2\xi\right)\right)}{M^4}$$

$$= \frac{1}{2}\left[\frac{1}{4}+\frac{9}{4}\right]-1=\frac{1}{4}$$

(269)

which is what we set out to prove.

To summarize this appendix, we have used stochastic simulations to show that

$$\rho_{n,k} = \begin{cases} 1 & n=k \\ \rho_1(\chi) & |n-k|=1 \\ 0 & |n-k|>1 \end{cases}$$ where $|\rho_1(\chi)|\le 0.3$. Furthermore, we used heuristic and

mathematical derivations to justify the asymptotic values of $\rho_1(\chi)$, namely that

$\rho_1(\chi)\xrightarrow{\chi\to 0}0$ and $\rho_1(\chi)\xrightarrow{\chi\to\infty}0.25$.

VDM
Verlag
Dr. Müller

Wissenschaftlicher Buchverlag bietet

kostenfreie

Publikation

von

wissenschaftlichen Arbeiten

Diplomarbeiten, Magisterarbeiten, Master und Bachelor Theses
sowie Dissertationen, Habilitationen und wissenschaftliche Monographien

Sie verfügen über eine wissenschaftliche Abschlußarbeit zu aktuellen oder zeitlosen
Fragestellungen, die hohen inhaltlichen und formalen Ansprüchen genügt,
und haben **Interesse an einer honorarvergüteten Publikation**?

Dann senden Sie bitte erste Informationen über Ihre Arbeit per Email
an info@vdm-verlag.de. Unser Außenlektorat meldet sich umgehend bei Ihnen.

VDM Verlag Dr. Müller Aktiengesellschaft & Co. KG
Dudweiler Landstraße 125a
D - 66123 Saarbrücken

www.vdm-verlag.de